A Fish Ate My Homework

A Beginner's Guide to Fly Fishing

Rick Robinson and Wade DeHate

Headline Books, Inc.
Terra Alta, WV

A Fish Ate My Homework

by Rick Robinson and Wade DeHate

copyright ©2022 Rick Robinson and Wade DeHate

All rights reserved. No part of this publication may be reproduced or transmitted in any other form or for any means, electronic or mechanical, including photocopy, recording or any information storage system, without written permission from Headline Books, Inc.

To order additional copies of this book or for book publishing information, or to contact the author:

Headline Books, Inc.
P.O. Box 52
Terra Alta, WV 26764
www.HeadlineBooks.com

Tel: 304-789-3001
Email: mybook@headlinebooks.com

Cover photo by Manni Svensson

ISBN 13: 9781951556785

Library of Congress Control Number: 2021950534

PRINTED IN THE UNITED STATES OF AMERICA

To Linda Robinson and Robin DeHate

The chapters in this book were written by Rick Robinson and Wade DeHate. But once the authors were finished with a given chapter or topic, they invited fellow fishers (and a few non-fishers) to contribute. Thanks to those taking the time to do so.

Please note the authors who have written full-blown essays for this book have given Headline Books a limited license to use their works for this book only but have retained their copyright. Those essays are noted with the copyright symbol ©. For those contributors, please contact them directly for the rights to reproduce their essays.

A portion of the proceeds from book sales will be donated to Trout Unlimited.

If you, your store, or your club would like to buy bulk copies for resale or gifts, Headline Books will co-brand the front cover with your logo or name. For more information, contact cathy@headlinebooks.com.

In memory of one of the great fly fishers of all time, wildlife artist W.D. "Bill" Gaither (1942 - 2016), seen here in his only self-portrait.

"Folks can tout their favorite bone fishing stores from San Andros, or speak of brook trout on dry fly, but few fish have the grit and stamina to equal a stream smallmouth bass. My favorite fish on this earth."

—Bill Gaither
May 26, 2013

"Never leave fish to find fish."
—Moses, 1200 B.C.
(or thereabouts)

Introduction

"I normally write political thrillers, but I've come to realize I like fish more than politicians."
—Rick Robinson

I started writing this book in the middle of the 2020 COVID pandemic. With little else to do except occasional trips to fishing waters, I kept writing in 2021. It is not as if I found inspiration in the middle of a horrific health crisis and a tumultuous world economy. The world spinning wildly out of control is only inspiration for headline writers and folk singers. The reasons I started this particular project were: a) my publisher made me, and b) during the worst of these times, fishing became my number one mental refuge. Scottish novelist John Buchan once wrote "**The charm of fishing is that it is the pursuit of what is elusive but attainable, a perpetual series of occasions for hope.**" Over the last couple of years, we have desperately needed occasions for hope.

The possibility of being the recipient of a sneeze-induced, microscopic-sized, virus-laced particulate (which, by the way, ironically looks like Wilson from the movie *Cast Away*) has changed the lives of people across the globe.

Reactions to the pandemic have been mixed. The entire world seems to either be searching for protective masks or railing against a government conspiracy founded in a belief in masks leading to the downfall of Western Civilization as we know it. As for social distancing, people either go full-tilt bubble-boy or attend frat-boy keggers on the beach. As with politics these days, there is no middle ground on COVID either. And do not even get people started on vaccinations. I have survived frozen margarita headaches worse than the reactions people have experienced from uber cold COVID vaccinations.

One large group of people seeing no change in their daily outdoor routines are those of us who are fishers – ordinary folks lining the shores of lakes, ponds, brooks, streams, rivers, seas, and oceans in hopes of catching a glimpse of their favorite sport fish. Those who practice the sport of fishing practically invented the guidelines issued by the Centers for Disease Control and the World Health Organization. We have been practicing COVID procedures since Jesus stood on the shore of the Sea of Tiberias and yelled at Simon Peter to have his fishing buddies spread out and cast nets to the right side of the boat. As good of a guide as you could find, I follow His advice and always make my first cast to the right. An intricate part of fishing strategy, He and His Father are also the first people to blame when a trophy trout spits the bit. Remember, there is a Commandment against such declarations, even when fishing.

Throughout the pandemic, part of my near-daily routine was to spend an hour or so pestering fish. I have two small ponds within walking distance of my home and two lakes stocked with trout a short drive away. My coauthor, Wade DeHate, lives near a portion of the Cumberland River called Rainbow Run. During the COVID crisis, I have regularly visited these spots without any fear of infectious disease. I had no apprehension about doing so because fishers have been social distancing, masking, and washing our hands forever.

Rick Robinson and Wade DeHate

First and foremost, fishing is a sport specifically designed for social distancing. People purposefully spread out along the shore to keep their lines from getting tangled up. And at every pond, river, or creek, each has their favorite space where they fish and protect a golden "honey hole" like Pooh Bear protects ... well ... honey. Once you get to "your spot," you spread out a border wall of chairs and gear to protect a zone at least six feet wide. AT LEAST.

Truer words were never spoken than when lawyer and author John D. Voelker (who authored *Anatomy of a Murder* under the pen name of Robert Traver) declared, **"Most fishers swiftly learn that it's a pretty good rule never to show a favorite spot to any fisher you wouldn't trust with your wife."**

To add to my personal social distance protection, I mainly fly fish. The whipping back and forth of a small, barbed hook at the end of a fly line would cause most Marvel Comic superheroes to keep their distance, and DC Comic characters too if they could afford to go to the river. Fly fishers are particularly practiced in the art of social distancing. I have watched in amazement as men and women line up in a river or stream hot spot with almost the precise same distance separating them, each tossing nymphs and midges upstream in a ballet of untangled tippet – the very thin position of line at the end of your leader. This display of grace is the closest any of us fishers will ever get to the Bolshoi and, while there is less choreography, the solemn reverence is all the same.

Finally, just to ensure approaching virus spreaders keep their distance during the pandemic, I smoke cigars while I fish. I still do. I think it was Mark Twain who once said something to the effect he smoked cigars to keep mosquitos and most people at a distance. Author Nick Lyons makes my point better, **"I fish better with a lit cigar; some people fish better with talent."** I bet Nick and I smoke the same brand.

Ernest Hemingway once said, **"Somebody behind you, while you are fishing, is as bad as someone looking over your shoulder while you write a letter to your girl."** Hemingway was a chauvinist. The best fly fisher I know is someone's "girl." For Ernie's quip, I offer sincere apologies to Janine Young, wife of my

11

high school pal, Jim Young. His wife is tenacious on the river. You'll read their story later on in the book.

Last fall, I met a grey-haired lady on Wolf Creek in Kentucky, and we spoke briefly about what was biting and other life situations. "What the hell else am I going to do," she said, puffing on a menthol. "Go to the mall and shop for shoes?" Not being sure if there was a mall within 100 miles of where we stood, I moved her ahead of Margaret Thatcher on my list of most admired women.

But I digress.

So, for the most part, no one gets near smoky me wildly slinging a San Juan worm over and over again into my favorite hotspot.

Face protection became a big story during the pandemic. One government mandate got passed, and it lit up the internet like grumpy cat memes.

Yet, as for face protection, fishers have been wearing masks long before N-95 masks were trading hotter than Bitcoin. To charge us a premium, fishing apparel companies call these masks skin gators and sometimes generically referred to by the company name that makes the majority of them – Buffs. Structurally, buffs are ultraviolet ray blocking, 1980s tube-tops for your neck, mouth, and nostrils. Anglers buy buffs to keep the sun off their face and ears. Along with UV protection, they come with the added benefit of keeping particulate from someone who just sneezed either in or out, depending on your buff perspective.

The joke is on the government. Fishers were the pioneers of the greatest 2020 fashion trend before it was cool. The only real change was camo buffs were replaced by Lynyrd Skynyrd logo N95 masks.

The masks and distance do not make us anti-social. Masked and distanced fishers talk constantly. We just have no idea what anyone else around us is saying. I went fishing with my coauthor the day before writing this introduction. We were a good bit apart with our buffs covering our mouths, ears, and noses.

WADE: "I caught a bass on a blue lure."

ME: "No, I never smoked grass watching Ben Hur."

We both gave each other a big thumbs up to acknowledge the social interaction and not be embarrassed, yet we had no earthly idea what the other had just said. I highly recommend the use of buffs next time you argue with your spouse.

Finally, everyone told us to wash our pandemic spreading hands like obsessive-compulsive germaphobes. That, of course, was not a problem for fishers. Fishers wash their hands – A LOT. Have you ever smelled the hands of someone who has caught a fish with live worms on a hot summer day?

I went trout fishing on the Cumberland River one weekend in early fall, and my nephew cut up the day's catch of rainbow trout. Three days later, he was still trying to wash the stink off his hands. He is a teacher, and I suspect when his school resumed classes, some kid asked, "What's that smell?" Of course, the kid had a mask on, and it came out, "Umpf Ta Klall?" So, no one had any earthly idea what he asked anyway.

Once again, in 2020, fishers delivered us the greatest communication tool of our age – a mask allowing us to socialize without the pressure of understanding.

I took up fishing in the fourth grade. My dad took me to a 4-H day camp where there was a drawing for a 12-gauge shotgun. Much to my surprise and delight, I won. My father, who prior to having crippling arthritis, was quite the outdoorsman, was ecstatic. My over-protective mother did not share in our enthusiasm. Fearing my next target would be a neighbor (as opposed to a 4-H clay pigeon), my dad and I were sent to the sporting goods store to trade the shotgun for fishing gear.

We left the store and immediately went to a nearby pay lake. My dad strung up a cane pole and taught me how to put the worm on the end of the line. To finish the ensemble, he put a bobber the size of a softball on the line. "When you see the bobber move," Dad instructed me, "tug back real hard to set the hook." After some waiting, the bobber moved. I did as was instructed, yanked with all my might, and watched as my first bluegill went airborne over my head into the brush behind us. For his protection, Dad quickly upgraded me to a Zebco with a thumb release for casting and cranking bait.

From that day forward, some of the greatest memories of my youth were fishing with my father. It is not like we went every weekend, but I remember each time we hit the water. Getting up early for a breakfast of scrambled eggs and bacon, driving to the lake, wetting the first line, eating sandwiches Mom had made for us, falling asleep on the drive home are all memories etched in the reaches of my mind.

My compulsion with fly fishing came only in the last couple of years, and it suddenly changed my fishing mindset. A fly rod in my hand cleared my mind in a way I have trouble putting into words. Maybe that was the precise point. I even started tying my own flies. When the pandemic hit, the sport became my getaway. I never really cared too much about what I caught (look at pictures on my Facebook page of me holding fish and you'll see it's a lie, but it sounds good); I released everything anyway.

As COVID continued its destructive path, I kept fishing. Buying hooks and tying flies seemed cheaper therapy than engaging a shrink, and my gear came without having to sign a convoluted HIPAA disclosure. I revived my weary soul at various fishing spots – wearing my buff, protecting my turf, and washing my hands like a shovel supervisor at the manure factory.

And that's how this book was magically born. Well, not exactly magically. My publisher, Cathy Teets at Headline Books, tried to find me one day. Of course, I was fishing. When she finally caught up with me, the conversation turned to the fact she always wanted to publish a book about fly fishing. As I am a novice fly fisher who spends more time untying tangled tippet than actually catching fish, I scoffed at the idea.

Additionally, I was not sure what I could offer to the world of fly fishing, something not already expressed. Between *The Orvis Guide to Beginning Fly Fishing: 101 Tips for the Absolute Beginner* and *A River Runs Through It and Other Stories,* I wasn't sure there was much else for me to say. Luckily for me, I had a secret weapon. I had Wade DeHate.

Retired Hillsborough County, Florida Fire Chief Wade DeHate, a much better fly fisher than I, made the concept for this book easy. The Chief had taken me on two guided fishing expeditions – one on the Colorado River and the other on Kentucky's Cumberland River. "People don't necessarily fly fish to catch fish," the Chief said. "People fly fish for the solitude of the experience. Catching a fish is merely icing on the cake."

And in a snap of a fly line, we had a book. Luckily for me, the COVID references will endure as long as new strains of the virus will likely continue to evolve until there will be one named after your street.

While we'll offer a few tips here and there, this is not a book of fly fishing technique. G. E. M. Skues, the British lawyer, author and inventor of the nymph fly, once declared, "**No art could be learned from a book, and fly fishing is no exception.**" I concur.

This is a book to help the reader understand why so many otherwise perfectly sane men and women can stand waist-deep in cold water, changing out a new fly every ten minutes and remain perfectly content to catch nothing but the moment. We offer an explanation about people you know and love who actually PREFER to spend their free hours dressed like characters from a fishing reality show and smelling like a day-old, defrosted box of fish sticks. This book is about how we find peace.

This book is meant to get you out of your chair and go wet a line.

Thus, I begin our book with this philosophical advice:
"Many men go fishing all of their lives without knowing that it is not fish they are after."
—Henry David Thoreau

And Hank should know. He never caught COVID on Walden Pond.

—Rick Robinson
Crestview Hills, Kentucky
January 1, 2021

Why do we fish?

"Beginners fish because it's easy to practice. The fish eat your homework."

—Wade G. DeHate

I do not think I'm alone when I say going fishing is an effort to simplify things in your life for a little while. Fishing is a timeout from life's commitments, enjoying a peaceful atmosphere, and maybe, just maybe, involve the art of catching a fish.

Many of us can probably remember an early fishing experience involving a cane pole. The cane pole is a simple device with no reel, a fixed length of line, a hook, and maybe a strike indicator… excuse me, a bobber. People have been angling for centuries with a hook and line with some type of organic bait or a deceptive artificial offering dangling from the end of the line. Historians give many examples of people trying and succeeding in catching fish that may or may not have been attached to some sort of a stick.

Interestingly enough, incorporating a reel to aid in retrieving a hooked fish is relatively new in the period humans have been fishing. If you are new to fly fishing, it may surprise you to discover just how often you will catch and retrieve fish without much use of the reel. If you keep reading, you will hear me sometimes comment the reel is "just a nice place to store line."

It is my opinion fly fishing is just not as complicated as it's sometimes portrayed. Sure, entomology is hard, but we have researchers to explain it. Some people will fall hook, line, and sinker for my uncomplicated comment, but it's true. The goal of fly fishing is the same as any other type of fishing, but for sure, fly fishing will take more effort than an occasional glance at a bobber and a worm on a hook. If you are already a spin or baitcasting angler, adding another successful method of catching fish while enjoying the great outdoors is already in your DNA. Do not be intimidated by the appearance of fly fishing as too complex. The fly rod will afford another tool to do what you like to do: go fishing. Fresh or saltwater venue, trout, bass, bluegill, or bonefish, go for it. We tend to relate to fly fishing for trout because it is what we like. Still, we will not cast pious judgment on other (legal) types of angling and do not subscribe to the snooty fly fisher mentality some folks are prone to exhibit.

I love to fish and have fished fresh and saltwater from Alaska to Scotland and south to the Virgin Islands. Ultimately rivers and streams are my favorites, and trout fishing for rainbows, brownies, and brookies is a beauty unique to fly fishing. With trout fishing in mind, you will see a few quotes from well-respected authors like Norman Maclean in this text. We credit Maclean for his observations because they are sincere, poignant, and reinforce the fly fishing premise we are attempting to explain, but of course, he does it so much better. Strangely enough, Maclean's *A River Runs Through It and Other Stories* novella was his only book published during his lifetime and is probably one of the best examples of combining the technical aspects (fly fishing) with a great story. It helps to get an idea of the wit and fly fishing wisdom of Maclean if you listen to the June 8, 1976, Chicago History Museum recording of the interview conducted by famous broadcaster, author, historian, and actor Studs Terkel:

> **Terkel**: "Your father and you, since you were Presbyterians and Scots, Izaak Walton had rated low in your pantheon."
> **Maclean**: "My father says he was an Episcopalian and a bait fisher and not, therefore, a respectable writer."

In deference to the above exchange, I am guilty of both the religion and – in my younger days – fishing with a coffee can full of freshly dug worms, but I am actively seeking reformation of my mediocre writing skills. If I was to fetter it down to three things best about getting into fly fishing:

- Fly fishing opens a whole new method of catching fish.
- The fishing locations (especially trout) are some of the most scenic you can imagine.
- It is another invaluable venue for a peaceful outdoor experience with loved ones.

You will see multiple recommendations in this text to hire a guide (or go with an experienced friend or relative) or take a class at a fly shop to get started in fly fishing. This one simple step will eliminate 90% of the aura of complexity and quickly get a "fish on" your fly rod.

Here in these pages, we have attempted to assemble a myriad of experiences and expertise focused in an easy-to-read format to help the novice and the experienced fly fisher add more to their toolbox. We all stand on the shoulders of women and men before us, and we encourage you to digest the history of where we and others have been and where we are going with fly fishing, conservation efforts, and enjoying these beautiful resources for our next generation of fly fishers.

I honestly feel the solitude, camaraderie, and beauty around us is paramount, and catching can be uncommon but still fulfill the need.

Go Fly Fishing!

—Wade G. DeHate
Somewhere in Kentucky on the Cumberland River
January 1, 2021

1

Is Fly Fishing Really That Hard?

"If I fished only to capture fish, my fishing trips would have ended long ago."
—Zane Grey

The Perception of Fly Fishing

Your first big decision is to take up the sport of fly fishing. We hope by purchasing this book, you have made the commitment. Do not treat it lightly. It is a big deal simply to make the mental decision to tr.

Like taking up tennis, you will need some basic equipment.

Like taking up golf, you will need some basic instruction.

Like taking up skeet, you will need some basic hand-to-eye coordination.

But the similarities to every other individual sport in which you may have participated end there. It is time to change your mindset.

Serve up the perfect ace just inside the line, and you pump your fist in triumph.

Hit a driver on the screws and stand over a golf ball at "Position A" on a dogleg left, and your heart races.

Blow up a clay pigeon, and it is a mental rush.

For fly fishing, get those ideas of instant gratification out of your head.

The first thing you learn about fly fishing, compared to any other individual sport, is that you are rarely rewarded with a fish following an ideal cast. For some reason known but to God, fish do not always choose to cooperate with us.

Tossing your fly into the perfect seam of rushing cold water and waiting as a school of unamused trout watch it pass by … well, makes you understand it is time to think differently about the rewards of individual sports. The perfect cast becomes the ace serve … the hole in one … the dead clay pigeon. The fish on the end of the line is a secondary level of fly fishing satisfaction we'll get to later.

The Reality of Fly Fishing

Some of our best days of fishing resulted in never getting a fish to the net. In *A Fishing Life,* author William Tapply noted, **"I've gone fishing thousands of times in my life, and I have never once felt unlucky or poorly paid for those hours on the water."**

Many experts say the first thing you should do when approaching a potential fly fishing venue is to read the water for where the fish might be hiding. Those people are quite wrong. The first thing you do is clear your head. Remember, the worst day of fishing is far better than the best day at work.

> *"Fishing is the chance to wash one's soul with pure air. It brings meekness and inspiration, reduces our egoism, soothes our troubles, and shames our wickedness. It is discipline in the equality of men – for all men are equal before fish."*
> —President Herbert Hoover

Herbert Hoover was a horrible president. The confusion exhibited in this otherwise eloquent quote shows why. Fishing is about cleansing the soul. It is true the fisher will at times discover there is no more sobering experience than being humbled by a fish. At those times, we look to our surroundings and listen to the sounds of chirping birds. Maybe Hoover was so inept at handling the stock market crash of 1929 because he considered fish as the end to the means of fishing. His saving grace was declaring God did not count fishing against our allotted time on earth.

Harry Middleton points out the fallacy of Hoover's well-crafted point of view. **"Fishing is not an escape from life, but often a deeper immersion into it..."**

History is filled with great people who found their refuge in fishing. Washington Irving, who wrote *The Legend of Sleepy Hollow* and *Rip Van Winkle,* enjoyed wetting a line. **"There is certainly something in angling that tends to produce a serenity of the mind,"** he once professed. And this realization comes from a guy who envisioned headless horsemen tossing flaming pumpkins and little men with weird names. David Stuver, author of *Familiar Waters: a lifetime of fishing in Montana*, was more precise in his assessment, **"...if catching fish is your only objective, you are either new to the game or too narrowly focused on measurable results."**

Being alone on a river or stream is hard to describe. Yet, John D. Voelker summed it up so eloquently, **"Fly fishing is a magic way to recapture the rapture of solitude without the pangs of loneliness."**

Subliminal Messages

There are several likely reasons people decide to take up fly fishing. The most common reason for interest in the sport is a family member or friend got them hooked (apologies for the pun) years earlier. Many people have fished all their lives, but never with a fly rod in their hand, but have a memory of somewhere or something driving the desire to learn. We are often approached (especially when fly fishing at a lake or pond) by others who want to know more about fly fishing. The conversation almost always involves watching someone they knew, loved, and respected slinging a fly. Often, they reappear a few days later, proudly displaying a 1969 Sears and Roebuck, Ted Williams Model 535.30454 Light Action 8' Fiberglass Fly Rod and a handful of flies looking old enough for Ted to have tied them himself.

"This belonged to my (insert old dead relative here)," they will boast. "I know you come here a lot and just knew you would like to see it." As they pull the crumbling line from the antique reel, their eyes light up like children on Christmas morning recalling stories of watching a special someone fish with the equipment they gently grip.

As a side note, Ted Williams was not only the greatest MLB hitter ever (fact – not opinion), he was a Hall of Fame Fly Fisher.

But, as we will often do when speaking of fishing, we digress. You can expect this a lot from two guys who repeatedly remind each other to put the plugs in the boat.

The folks with gear in their garage or a faint memory in their brain generally do not need to be sold on fly fishing. They already have it in their blood. They just need a little encouragement to pique their interest and a little schooling to keep them from snapping their rod into fifteen small pieces on their first outing. Some of my favorite fishing days are seeing these same people reappear with refurbished gear, replenished line, and a mayfly not looking like it was hatched in some bygone prehistoric era. A panfish caught on a gold bead San Juan worm becomes a trophy fish of forgotten memories.

It is then, and only then, the old fisher – reincarnated as a new fly fisher – understands the words of outdoor writer Bob Lawless: "**If I'm not going to catch anything, then I'd rather not catch anything on flies."**

The second group of people take up fly fishing because they have watched or read Norman Maclean's *A River Runs Through It and Other Stories.*

These could be folks who have quit the local country club and are looking for something more out of sport than cursing at yet another bad shot before tossing their entire golf bag into the lake guarding the green of the signature final hole. This is where the mental reset comes into play.

Norman Maclean is brilliant and the voice for a generation of fly fishers. He spoke of the sport as a religious experience and being haunted by rivers cut by generations of raindrops.

Of course, there is a huge difference between craving the

emotions elicited by Maclean's words and the beginning fly fisher's first actual attempt at fishing. As comedian Steven Wright properly points out, **"There is a fine line between fishing and standing on the shore looking like an idiot."**

The questions for these people taking up fly fishing are: "How do I find that line? Should I put my barrel of used drivers up for sale on e-Bay? How do I know if I'll even like this sport?" Walking through the fishing section of a sporting goods store, you will suddenly be reminded of all the little-used exercise equipment you have purchased over the years. "Will this new fly rod be another unused item leaning against it?"

> *"I know I'm a long way from greatness, but I am beginning to come at it in my own way. I can go through the basic motions pretty well, don't rely quite as religiously on specific fly patterns as I once did, have worked out ways of compensating for some of my most egregious weaknesses, and have come to count heavily on timing because it's a hell of a lot easier to catch fish when the fish are biting."*
>
> —John Gierach (2010),
> *Standing in a River Waving a Stick*

Getting a Good Start

For the novice seeking a new sporting experience, we suggest a slow boat to nowhere. Literally.

Find the best fly fishing guide in your area and hire them to take you out for a day of angling. Any type of fishing, whether the venue is saltwater or fresh, lake or river, ocean or flats, you can fly fish it. Let the guide know you are at best a novice and at worst someone "standing on the shore looking like an idiot." As a good guide knows their client, you probably will not have to explain, but telling the guide yourself will make you feel a bit better about your fishing inadequacies. Guides come in all shapes and sizes, male and female, young and old. Leave your preconceived notions at home, and use the online reviews and personal references when choosing a guide.

The guide will bring all the equipment, tie all the knots, and give you some basic instruction on casting. For the rest of the day, do what the guide tells you. For this trip, you do not have to worry about anything else we write about in this book. Simply take in the experience. This should remove all doubt about the plethora of knowledge, technique, and advice you will receive. Guides are ideal for any level of fly fisher but particularly useful for the newcomer.

Brandon Wade of Cumberland Drifters (you can find him on Facebook) is one of the finest guides we have ever encountered and spends his days working the Cumberland River in Kentucky. Look for the "Fly Fishing for Trout on the Cumberland River from a Drift Boat" in our references section. We asked Brandon to list the five things novices should expect from their first guided fly fishing experience. Here is what he told us.

1. If you're brand-new to fly fishing, then the playing field is pretty flat, and most people (men and women) will *listen to the guidance of their guide* (redundancy intended).
   - Takeaway: **Listen to your guide.**
2. Teaching women has been experientially observed to be easier than teaching men. Women inherently "listen better" to instruction, and men struggle more with taking instruction, especially while learning a historically male-dominated sport of fishing.
   - Takeaway: **Listen to your guide.**
3. The number one frustration for any guide is their client does not utilize the guide's experience on their home water! Give great value to the fact the guide you select lives either on or near the water you are about to fish and fishes it regularly at different times of the year. They know the "honey holes" on the trip and want you to be properly rigged for each spot as you fish.
   - Takeaway: **Listen to your guide.**
4. If you're an experienced fly fisher – great! It will expedite getting into fishable water quicker, allow for faster adjustments to current conditions, and ensure quick advantage is taken of current presentation opportunities.

As an example, Brandon recounted a trout fishing guided trip for a married couple. The male a very experienced fly fisher (fly shop owner) and was accompanied his novice fisher wife. The fly shop owner immediately clipped off Brandon's pre-rig of the proper flies expected to be producing on this water at this particular time of year and insisted on using his own selections. At the end of the day, the result was one or two fish for the experienced fly shop owner, and his wife lost count of the many fish she brought to the boat.
- Takeaway: **Listen to your guide**.
5. Brandon recounted several other examples where friends and especially family members would not listen to obvious helpful hints or instruction, no matter how tactfully given. Sometimes the old spinning outfit works well, but other times only the fly rod will produce success.
- Takeaway: **Listen to your guide**.

Perhaps it's pride or stubbornness closing the experienced fishers' ears to advice. Or some are just convinced their method is better because it worked so well last week on their favorite spot. The justification for not following your guide's advice will become readily apparent within an hour into the trip. Your mission, should you choose to accept it, is to acquiesce to the help readily at hand because remember you paid for the trip. Besides, your spouse is gonna rub it in later and bring it up often.

Any guesses to the recommended takeaway from this segment?

Right – **Listen to your guide**.

Still Committed to Fly Fishing?

If your slow guided drift down a cold river is an effortless journey into the depths of your very soul (and **you listened to your guide**), proceed to Chapter Two. If, instead, at the end, you are unamused and do not see the point, go fish your golf clubs out of the lake on the 18th hole and regift this book to a friend.

You will not be alone. Author John Steinbeck was not a fan. "**It has always been my private conviction that any man who pits his intelligence against a fish and loses has it coming,**" opined Steinbeck. And in his book *The Philosophical Fisher*, Harold F. Blaisdell jokes, "**All the romance of trout fishing exists in the mind of the angler and is in no way shared by the fish.**"

While there may be many many reasons people take up fly fishing, there are reasons that do not apply to why we fish. In particular, we are not "meat fishers." As we will note later in the book, we do not keep and eat our prey for the most part. Usually, we keep fish only when they may not survive due to high water temps or injury, but this is rare.

Do not let the fear of mandatory harvesting prevent you from taking up fishing. It just ain't so.

Once you are hooked, there will likely be a long list of other reasons you continue. Good for you. Experiencing the joy of fishing is what the sport and this book are all about. Former British Prime Minister Tony Blair once said, "**Some go to church and think about fishing, others go fishing and think about God.**"

For those hooked from the first cast, a note about telling folks about the fish you caught (or nearly caught).

No one can figure out who said, "**Nothing grows faster than a fish from when it bites until it gets away.**" No one admits to saying this because they will have had to admit to the factuality of the statement. As for fish stories, Mark Twain advised, "**Do not tell fish stories where the people know you… but particularly, don't tell them where they know the fish.**"

The Saltwater Aspect

Many people think of river and lake venues when fly fishing is discussed. Visions of rainbow, brown, cutthroat, or brook trout, and bass and bluegill or pike and muskie dance in their heads. If you are a coastal resident or want to travel to the seashore to fish, a saltwater venue may also be a great choice for a fly fisher. Both authors have lived in coastal areas of Florida, and Wade grew up in the Tampa Bay area, home to some of the best inshore fishing anywhere. The west coast of Florida is home to outstanding redfish, snook, tarpon, and sea trout fishing. Be forewarned: Saltwater fly fishing is commonly performed with different tactics, procedures, and heavier-weight rods and reels than freshwater venues. The authors recommend first mastering fly fishing fundamentals. After gaining decent casting proficiency with lightweight tackle, move into saltwater involving greater casting distances and windy conditions open elements commonly provide. More on saltwater later with equipment selections and fly presentations.

A Fish Ate My Homework

In each chapter, we solicited the knowledge and experience from many of our fellow fly fishers from around the world on the chapter subject matter. Consider using "A LITTLE HELP FROM OUR FRIENDS" as you proceed in your fly fishing journey.

AND NOW … A LITTLE HELP FROM OUR FRIENDS

I have been fishing for 55 years: bass, catfish, and fly. I can tell you that wherever you go fly fishing, you will meet more and friendlier people. Everyone talks to the guy fly fishing, not so much the bass or catfish guys. So fly fish if you can; it's a lot more fun even when you don't catch fish.

—Charles Exline
Verona, Mississippi

A Fish Always Rises
By Spencer Durrant ©

A version of this story originally ran in Sporting Classics Daily July 9, 2018

"You don't want a job in fishing," Tom Rosenbauer once told me. "You want a day job. Something that'll actually put food on the table."

I wanted to prove Tom wrong, but back when we had this conversation, all I had to my name was a dozen fly rods, a freezer full of fish and wild game, and a bank balance that reflected the cost of traveling the West in a pickup truck. I had the world right where I wanted it, or so I thought.

Tom wasn't trying to discourage me from pursuing my passions; he was just being honest. It wouldn't matter if my freezer or bank account emptied first because either situation would leave me looking for work in an industry that's not known for its wages. Sure, you get an office with a view, but as the old saying goes, you can't eat the scenery.

That's why, while waiting for our food at the Arby's in Evanston, Wyoming, I peppered John Gierach with question after question on how he'd made a living. Between him and Tom, I'd been lucky enough to talk with the two most widely-read fly fishing writers in the world. John's advice was just as blunt as Tom's.

John graciously put up with my questions, and the whole conversation put a smile on Bob White's face. I have a feeling he's sat courtside for many a similar discussion over the years he and John have been friends. For the next week, the three of us would be sharing drift boats, car rides, and meals, so I reckoned I ought to leave the week with a bonafide plan for success in fly fishing.

Just as we all refilled our drinks for the last leg of the drive, John piped up with a final piece of advice.

"I've only known you for a few hours," he said, "But best I can tell is your problem is being too damn impatient. It doesn't happen overnight."

With that, he walked outside, lit a cigarette, and waited while I gassed up the car.

John was absolutely right. I proved him more so when, the following morning, I missed the first three fish rising to my dry fly because I set the hook too quickly.

It was late March, a bit too blustery for the blue-wings to really take off, but not so bad as to put them down entirely. We had the whole week in front of us for fishing, including two days of floating the Green River below Flaming Gorge, in Utah. This section of the river is world-famous, and its blue-wing hatches are the stuff dreams are made of. Fly fishers' dreams, at any rate.

On the first day of the trip, I proposed that we fish a small, unremarkable creek a few hours away from the Green. I knew John's preference for small water, and Bob is one of the best sports I've ever met. He would've been happy fishing a community pond with stocked rainbows. Not because he's not a good angler – he's absolutely outstanding and a better human to boot – but because he just loves fishing.

The creek we fished is a sagebrush-infested, snake-filled, sweltering oven of a place, even in March, but it presents a good chance for a trout over 25 inches. Any time you can catch trout that big from a creek that doesn't get wider than 20 feet in most places, you do it.

I was the de facto guide since I knew the area and had set up the trip. Bob hadn't fished in Utah before, and John rarely visits from neighboring Colorado. So, I set them up in a run I've had good luck in and sat back to watch them fish. I was completely unqualified to stand between those two men and tell them where to fish. John's writing and Bob's paintings have shaped fly fishing into what it is today, and without fly fishing, there's a very real chance I wouldn't be around to tell this story. Acting as a guide for these two was the most diminutive I've ever felt, and I've covered NBA and college basketball since I was 18. I'm used to being the small man in the room, but only literally.

Bob insisted John fish first, and John didn't need any encouragement. He lit a cigarette, pulled out a fly box, and looked over the rim of his glasses. A Royal Wulff was the fly of choice.

Nothing John did was fast or frenzied. He'd traveled hundreds of miles to be here, and it was his first real fishing trip of the year. Most anyone else I know would've been a dozen casts deep by the time John dressed his fly with floatant. As this book is for beginners, we should note that floatant is a combination of a wax and carrier in a liquid state, intended to keep dry flies from absorbing water and to keep the fly floating.

He hooked a fish a few casts into the day, turning to me and grinning. Through the bushy white beard and dark sunglasses, I saw the same look of excitement that plasters my face when catching fish.

"It never gets old," John said, as much to himself as to me.

Bob followed suit, though I could tell his decades of guiding in Alaska hadn't been spent on brush-choked streams. He caught a few more trees than fish, something he laughed off later that night at dinner. Throughout the week, I was continually impressed with how neither John nor Bob took anything too seriously. We were on one of the world's best tailwaters, but the conversations at mealtimes and during the car rides between launching and loading drift boats centered around Bob's wife and kids and John's friend who owns a fishing lodge in Labrador. We only ever talked about fishing while we were on the water, and even then, there wasn't an overhanging worry to show John and Bob a good time or make sure they got into fish. I have a feeling they would've been fine getting skunked because they both knew what I had yet to learn at the time.

On the last day of our trip, we stopped for lunch on a sandbar in the A-Section of the Green River. Charles Card, a guide with Spinner Fall in Dutch John, Utah, and my friend Ryan Kelly (a retired guide-turned-photography guru) got lunch ready while John, Bob, and I took in the scenery.

"This never gets old," I said. John cracked a smile. Bob did too.

That's when it clicked. That's when I understood how John and Bob could be so laid-back, so relaxed, and so happy in the moment. As they can both attest, there's a difference between hard work paying off in the form of a finished book or gorgeous painting and trying to force either into existence. Or, as John implied that first night at Arby's, there's as much to enjoy while patiently waiting for things to work out as there is when everything comes full circle.

Fly fishing isn't the same as it was when John started writing and Bob started painting. Both men are hugely successful, though, because they're at the top of their respective games. There's less money in the industry these days, and it's more competitive than ever, but people like John and Bob stick around not out of sentimentality but because they've been patient enough to see things through.

In an era where everyone's a "brand ambassador" and thinks their Instagram posts are worth a full-time salary, I reckon that lesson's been lost. John worked odd jobs for years before he was able to dedicate himself full-time to writing. Bob's story about how he became an artist included stints in two different careers.

Their success took time, sacrifice, and an understanding that having fun is worth more than what most "real" jobs pay anyways.

Spencer Durrant is a fishing writer, guide, and bamboo rod builder from Utah. He's the Owner/Lead Guide at The Utah Fly Fishing Company, the News Editor for MidCurrent, and a columnist for Hatch Magazine. Connect with him on Instagram/Twitter, @Spencer_Durrant.

All of us have moments in life when we stop and ask ourselves, "How on earth did I get here?" My fly fishing has caused me to ask that question several times, and the answer is a memory of the day when the sport captivated my interest. My wife and I had made the journey to Yellowstone National Park in 1996, and we were amazed at the abundance of wildlife and the intriguing geological formations. We elected to pull into a small parking area to have lunch, and it featured a rock outcropping that overlooked a portion of the Yellowstone River. I sat upon a rocky perch and watched a couple of fly fishers down below as they carefully worked the pockets behind a couple of boulders. As I observed them, I found myself more and more fascinated with the art of the cast and the anticipation of a strike from fish that only I could see. I could have stayed there watching all day, but a seed had been planted. Curiosity and desire grew from that moment. I purchased a book on fly fishing at one of the visitor centers the following day, and I read it cover to cover and several times after that. It would be several years later before my dream became a reality, but that did not stop me from reading and learning and preparing.

I have been a fly fisher for 19 years now, and looking back over the years, I have come to realize why the sport has been such an ideal match for me. By training, I am a biologist. I majored in biology, and I have always been at home in the great outdoors. There is also an artistic side of me, and I spend time each year carving decoys and making fly fishing nets and have always enjoyed drawing. I work as a Registered Nurse, a profession described as both an art and a science. So fly fishing for me is as much about the ecology of fish and their preferred foods as it is in the art of the cast and the inherent grace of the sport.

Fly fishing is so much more than catching a fish, although it sure is fun when you do! It is a sport that has taken me to beautiful places and wonderful waters. It is an endless world of learning, and there is a kinship among those who are hopelessly lost in that world and a quiet understanding between us. We all know why we fly fish without saying a word. When we do speak to one another, we often talk quietly because we all revere the peace and tranquility of the streams we so enjoy. We might be strangers, but we are indeed connected by a shared love of wild places and the much-needed break from the busy world in which we live.

—John Munch
Owatonna, Minnesota

As with a lot of men, we have had dreams for many years of being able to fly fish. I finally had the time to take up the sport. Very fortunate for me, I got introduced to a couple of guys on the river one day, and they were so incredibly kind to offer some of their time to teach me our local waters and the proper methods to fish them. I cannot tell you how much this helped me advance in a fairly short time frame. These two guys, now good friends, have been so kind and generous that it is hard to say thank you enough.

—Leo Poggione
Reno, Nevada

A Fish Ate My Homework

Stephens Court – A Fishing Story
By Peter Collin © 2019

It was tucked behind this row of house trailers. You'd get there by parking at the dead end of a nondescript subdivision. From there, a few steps down the path the locals kept mowed would bring you to a 1000-foot stretch of some of the finest fly fishing in western New York.

It was formed by an Army Corps of Engineers project meant to reduce bank erosion. An oxbow in the Genesee River made it eat its way into the outside bend, threatening Route 19 and all the homes alongside it. The newly dug channel split the river's flow, so the original water's path became a mere side channel, removing the threat to infrastructure. There was a happy consequence to dividing the river like that. See, the Genesee River has always been *just this close* to being a blue-ribbon trout stream. There are about 30 miles of gravelly riffles and runs that get stocked with trout and have different fly hatches. But trout are a cold-water species, and you can't always rely on the Genesee to run cold. In a hot, droughty year, most of these stocked fish won't survive until fall. We are told by the DEC that there's no natural reproduction on the Genny. But every so often, you'll get a cold, rainy summer that lets the trout survive into the following year, where they live that much longer and get that much bigger. I've been around when we got *two* chilly, damp summers in a row. The fishing we got after that was the kind around which you could form a religion.

I took my buddy, Jay, the fishing guide, to the Genesee during such a top-notch year. He released a fish, stared at the pool full of rolling backs of browns and rainbows still porpoising, and said,

"A guy would travel *hundreds* of miles to fish here."

This made me a little uneasy. There are fishing spots in New York ruined by publicity, and I grew to be protective of my home river. Of course, by the dog days of that summer, the sun shone hot, the water level dropped, and that year's crop of trout went on to feed the herons, minks, and mergansers. The next spring, we were back to normal with mostly cookie-cutter stockers.

The thing about Stephens Court was that at the head of the first pool was a trickle of a spring. It bubbled up out of the gravel, delivering clear cold water into a slow deep stretch. A dribble like that wouldn't make much difference to the entire flow of the Genesee River, but since the side channel carried less than half the cubic feet per second, it could be cooled enough for the trout to survive through the hottest summers. Fish in Stephens Court were predictably present, plump, and happy.

Just knowing about Stephens Court's stretch to the Genesee River made you a de facto member of the most secretive, exclusive fly fishing clubs of that time. The day that John Miller first brought me there, he said in his most serious voice: "You are not to tell anybody about this place." He then named specific guys that I wasn't to tell – like Mark, who liked to talk so much that he wouldn't keep quiet, or Jay, the fishing guide who would have a pair of paying clients parked there from May until August. The entire time I fished there, I almost never ran into a stranger, just John, Merle, Bob, or Bill.

A guy prospecting the river from above or below by foot or kayak would be far more likely to follow the main channel, oblivious that he was sidestepping a fisher's Nirvana. Even if somebody were to make his way down the Hallowed Half, he might not realize what he was seeing. The water at Stephens Court was calm and the fish spooky. And if he didn't tiptoe through the head-high canary grass on the bank, or wade at a smooth, slow, Tai Chi like pace, your quarry would dart to an undercut bank or the belly of the pool. You may never see them or know they were there.

I kept my word to John Miller, for the most part. The few buddies I brought there lived far away, so there wasn't much danger of them showing up without me. I could pile dozens of individual stories on top of Stephens Court involving explosive mayfly hatches, super-selective risers at an evening spinner fall, my discovery of the LaFontaine caddis pupa making impossible fish easy, resident dogs getting into barking matches with mine from across the river, a velvety buck walking past me close enough to pet, a frightened turkey hovering and pooping over the river until he gained enough altitude to scram.

You can begin to feel possessive and protective of public water. That little, not even a quarter-mile stretch, just four miles downriver from the Wellsville sewage treatment plant, was a source of anticipation, adventure, and camaraderie for years.

Heraclitus said that no man can wade the same river twice. He probably meant that allegorically. But an aerial view of the Genesee shows how true that literally is. Any river is a dynamic environment, constantly rerouting its flow. Slow erosion or a quick diversion brought by a flood will shift the river's course. And what the Army Corps of Engineers giveth, Mother Nature can take away. Somewhere around 2006, I went there to discover that an early spring spate had dumped a gravel bar across the inlet to Stevens Court Run, cutting it off almost entirely. The magic waters that birthed our secret society had choked to a trickle you could jump across. Not willing to accept this, I went back there with a shovel and tried to dig the channel open by hand. The fresh gravel was loose and easily dug, but there was just far too much of it. I needed an Irish crew from the Erie Canal days to accomplish what was needed. Mind you this was a bit of guerrilla engineering. In New York, you need some serious permitting to divert a waterway. To go through the proper channels, I would have had to team with somebody like Trout Unlimited to lobby for the alteration, wait out the necessary environmental reviews, argue in a public forum – it could take years, and success would not be guaranteed. And, after all that, if the river were put back to where it was, it'd be known to all fishers far and wide, replete with a midstream ribbon-cutting ceremony in hip boots. The very obscurity that made Stephens Court what it was would be gone forever. I would just have to say goodbye.

So since those days, the Genesee River has changed, and maybe I have too. Insect hatches no longer produce the kind of fluttering clouds that draw swallows and waxwings. Most summers are now too hot and long to hold the trout over into the next year. My visits to the Genny are way less frequent and less bountiful. They're done as a tribute to times now past. I have moved on to boat fishing, first in the Finger Lakes and then in Lake Ontario. I found tremendous sport in these new places, of a completely different kind. Most lake trollers could never understand the sublime joy of getting a 16-inch brown trout to take a dry fly, or tying a perfect imitation of a stonefly you found on the stream, then fooling a fish into biting the next day.

The weeds grow tall on the old Stephens Court Run. John is gone. Merle is gone. Mark is gone. I turn elsewhere to find sport and adventure and experience them with different people now. Stephens Court exists in my memory, an example of discovering something that I had always dreamed about and finding its reality as wonderful as my imagination had supposed. I hope you all should be so lucky.

Authors' Note – A video by the same name as the article, produced and narrated by Peter, can be found on YouTube.

Depends on who you are. If fishing is about the fish, buy a guide. If fishing is about the fishing, you can, and should, do that alone.

—Tom "Gator" Gaither
Ludlow, Kentucky

Thirty years ago, I taught myself fly casting. As a beginner, I used WF line. For those new to the sport, WF stands for Weight Forward and refers to a type of fly line that has additional weight and thickness added to it in the first 10 yards of fly line, which allows the line to go out more easily. I watched the line as I cast. Kept it about 30 feet in length and watched as it followed the tip of the rod. Made a few forward and back casts. Then practiced laying it on the grass. Practice – practice – practice.

—Johnny Dennison
Lumberton, Texas

Why do I fly fish? **"Many men go fishing all their lives without knowing it is not fish they are after."** – Henry David Thoreau. All my adult life, people have frequently asked why I fish so much. The short answer is it's good for my soul, but the complete answer goes much deeper.

When I first became interested in fly fishing, it was because I thought it was a bit of an art form. How little did I realize that it was, and I would become addicted to this method of expressing myself. Standing in the middle of the water waving a stick around may not seem like much of an art to you, but we fly fishers like to think it is.

I have found it to be a way to bare my soul, commune with Mother Nature, and release all the pent-up frustrations from my work. Listening to the burble of a brook, watching insects hatch, seeing a wood duck family swim upstream, seeing a hawk take a squirrel from a tree branch nearby, hearing a sharp squeal then seeing a weasel come out of the rocks with a mouse in its mouth: These are things I have witnessed because I waded streams and fly fished. Fish don't care how rich I am or what my resume says. They don't criticize me or laugh at me (well, sometimes I think they do). Fly fishing helps me connect with nature and reminds me of what's important. Fly fishing makes me a better person, and most of all: It makes me happy!

—Joel Stansbury
Florence, Kentucky

2

WHAT FLY FISHING EQUIPMENT DO I NEED?

"My biggest worry is (when I'm dead) my wife will sell my fishing gear for what I said I paid for it."
—Koos Brandt

A Good Attitude, a Fishing License, and Time
It sounds a bit condescending but stay with us for a moment. The first and foremost item you need for fly fishing is a good, patient attitude. Essentially, you may need to redefine the meaning of success in the sport. We can assure you fishing success does not necessarily come out of your fly or tackle box (i.e., your wallet). It is perfectly fine (in fact, expected) for your technique to suck for a little while. Using the golf and tennis analogy again, keeping the stick in your hand improves muscle memory.

Just like hitting golf balls on the range improves muscle memory, being on the water improves the solace you feel being there. The motions become more natural and executed without even thinking too much. The reward may be the time lost in your thoughts. It may be teaching a child. Possibly, it is the desire to return sooner rather than later. Catching fish will come, but like with any resume, it takes a little education and experience to get the nod from any self-respecting fish. If everyone determined the impact of learning the fine art of fly fishing upon their soul, fishing stores would be back-ordered on fly rods and white woolly buggers.

After attitude, consider time as your second greatest asset. Novelist Thomas McGuane pointed out, **"Angling is extremely time-consuming. That's sort of the whole point."** Of course, while McGuane called fly fishing the Sport of Kings, he also declared, **"It's just what the deadbeat ordered."** Norman Maclean described fly fishing as an art. We believe both McGuane and Maclean.

Whether KING or DEADBEAT (we are not judgmental), let us start your education by putting the artist's tools in your hands.

After you go fly fishing with a guide, you will likely be juiced up and ready to buy some gear. A piece of friendly advice here – it is not necessary (or intelligent) to start out buying the most expensive items in the store. Every year, the Kentucky Department of Fish and Wildlife hosts a fly fishing rally for children. It is a blast to attend. The point is we have seen a four-year-old kid with a pink Hello Kitty rod and reel catch big fish.

You do not need the most expensive gear on the market. Even if it means a road trip, visit some independent fly-fishing shops near you. They are likely more invested in your success as a future customer. As you become more interested in the sport, you will discover you want new gear to fit different water situations.

Also, the legal aspect must be addressed. The folks at your state's Department of Fish and Wildlife work incredibly hard to make sure we can enjoy our chosen sport of fishing. Do not be a jerk. Buy a license for the fresh or saltwater venue (yes, they are often different even though they are geographically close). If required, get a trout stamp too. Know your state's rules and regulations. Put your license in a lanyard and be ready to show it to a local official when asked. They see everyone on the water and may even offer a tip or two on what flies are hot on any given day. Rick remembers a day when the county game warden encouraged him to move to a different part of a lake where fish seemed to be hitting.

The Gear List

The Rod

The first shopping trip for fly fishing equipment is seemingly a grouping of decisions with no apparent order of importance. This is not the case. The rod definitely comes first.

This singular decision involves an unlimited number of variables causing you to wake up in the middle of the night in a sweat as cold as a mountain-fed trout river. Should you buy a fiberglass rod, opt for a composite pole, or jump right into the "gold standard" – an expensive, handmade bamboo rod. What does the one X to ten X mean? And do not even get us started on the color of the line and whether or not it should be floating or sinking.

In his book, *Anatomy of a Fisher*, Robert Traver declared, "**...buying a fly rod in the average city store, that is, joining it up and safely waggling it a bit, is much like seeing a woman's arm protruding from a car window: all one can readily be sure of is that the window is open.**"

The purchase of the rod and reel is one of the few times when golf and fishing are similar. Every spring, pro shops and sporting goods stores are filled with golfers looking for a new putter and driver. The desire for a more accurate roll, an Nth degree straighter, and five yards longer is what keeps the golf equipment industry alive. And, while drivers may be exchanged for newer models, putters are kept like shares of Apple stock meant to enrich generations to come.

Such is the case with fly rods and reels. Assuming you followed our advice from Chapter One about going fly fishing with a guide, you are already a world ahead of the average schmuck standing in a sporting goods store looking at a wall full of rods.

Many fly-fishing teachers and authors advise you to follow the very basic "Rule of Fives" in making your initial choice. We are no different. If you are buying a rod for trout fishing, we recommend you start with a five weight mid-flex rod, a five weight fly reel and line, and 5X leader and tippet. This is a great foundation for your initial venture into the sport of fly fishing for trout, bass, or bluegill.

The X factor in fly line and tippet is a way to measure the width and strength of the leader you are purchasing. The packaging label will display an X Factor of 0 through 9 and give some basic information on how much weight it will hold. The X Factor is the equivalent to how much strength (in pound-weight) will be on a similar spin or bait casting fishing line, but the system is backwards. In other words, 0X line is the strongest with about 16lb breaking strength, while the 9X has only about 1.5lb breaking strength. This numbering system confuses most in the beginning but it is the system you inherited. Five X is a good place to start, because it puts you right in the middle. You can experiment with going up or down from there.

Author Reg Baird once said, **"The man who coined the phrase 'Money can't buy happiness' never bought himself a good fly rod!"**

As a general scale for selection of rods, reels, and fly line, consider the following:

Fly lines/rods/reels need to be matched together and come in sizes 1-12 weight (a higher number is a heavier line)

- 3-4-5 weight is average for most inland trout and bluegill-sized fish
- 6-7 weight for larger trout and bass
- 8-9 weight for pike, big bass, steelhead, and salmon, redfish and snook
- 10-12 weight for tarpon and similar heavyweight fighters
- Note: Saltwater fly fishing usually requires you to send flies out a good distance from the boat or shore, the fish are definitely bigger, and it's not uncommon to be out there on windy days. This requires 8-12 weight gear AND advanced casting proficiency to really be able to take advantage of the fishing opportunities presented in larger saltwater venues. We'll talk more about the need for double-haul casting in Chapter Three for effective reach to any saltwater fish.

If you have never been saltwater fishing before, and this is going to be the first saltwater fishing trip you have ever taken, our advice is:
- Hire a spin or bait-cast fishing guide
- Choose an *inshore* fishing trip for calmer water
- Expect a great time fishing
- Caveat: If it is summer, prepare for sunburn/windburn from the reflection off the water, as well as from above. Without fail, use hats/neck gators/gloves/polarized sunglasses. If it is spring or fall, bring a light jacket (yes, even in Florida). Even if you hail from the Yukon, you WILL get cold.
- Inshore fishing locations usually do not cause seasickness. If you happen to encounter some small but not too bumpy boating, many people have success using motion sickness patches.
- Lastly, after your first saltwater spin or baitcasting trip, think about how a fly fishing trip would be in those same conditions. It would be a whole new and different adventure. If you're already armed with decent fly-casting experience and the ability to apply it to the larger tackle/weight of saltwater, you will be in a much better position to enjoy the outing.

If fly fishing were a movie, the lead actor would be the fly rod (not Brad Pitt – an avid fly fisher, I'm sure he would agree). Fly fishing is designed around taking full advantage of everything the fly rod can offer, so try to find the best rod that will not break your bank. Try out different rods to find what casts well for YOU.

To get a bit more specific – considering everything else on the shopping list, as well – we suggest purchasing the best nine-foot rod your budget will allow. Your first fly rod and reel is far more than your initial connection to the sport. It becomes the measure against every rod and reel you will buy in the future. And, trust us, you will buy more.

Both authors have broken rods while either fishing or closing car doors. Purchase your rod from a company offering an unlimited long (or lifetime) guarantee. Every fly fisher will warn you: There will come a day when you need the warranty.

The Reel

If you are an existing spinning/crankbait fisher, or heaven forbid you go both ways (see "etiquette" section), there are some serious differences between fly fishing reels and other reels. Unlike a spinning or open-faced reel, a fly reel is, most of the time, simply *a nice place to store excess line* (fresh or saltwater).

There are many ways you might hook, fight, and land/net a fish. Many species of smaller fish (usually under 14 inches) can be caught with just your finger(s) applying drag/resistance on the fly line and then stripping line back with your free hand, bringing your fish to the net. If your fish does not threaten to overwhelm your leader and tippet strength and does not fight to the point of stripping more line from the reel, you may be able to fight the fish just by holding the line in one hand and the rod in the other, letting the coil of stripped-in line pile up at your feet.

When larger fish hit, the game is on, and they may strip out any slack line and "get on the reel" quickly. This is where your fly reel drag system comes into play. NEWS FLASH: The average fly reel does not offer the same drag feature as a spinning or baitcasting reel. We are saying that a spin or bait cast reel will allow you to attempt to reel in line while the fish is still pulling line out against the drag. Most fly reels are not equipped to operate in such a manner.

The ability to try and reel in line while the drag is being taken out is not possible with the common fly reel. This is one of the most significant transitions to fly reels for the average spin or baitcast fisher. As an example, if a nice fish is taking the line out on a fly reel, the drag will work where you set it to slow the run. Still, the retrieve knob will be spinning quickly, and if you grab the handle to try to be ready to wind in line (as you would on a spinning reel), you will stop the smooth performance of the drag operation and most likely overload the leader/tippet and break off the fish. Worse yet, some end up with banged-up knuckles and a lost fish. Both results are frustrating to any level of fly fisher.

This can happen with bigger rainbows and brown trout, but ask any fly fisher who pursues muskie, pike, steelhead, salmon, and many saltwater fish about trying to grab the knob on the fly reel as it "blurs around" in a circle as your best fish ever streaks away. If you try to stop a run on a big fish, bruised knuckles and a busted tippet are not uncommon results.

If you want to ensure this never happens to you, some fly reels are equipped with a couple of different types of "anti-reverse" features, allowing the drag to work without the knob spinning. If you get a bruiser on the line, the anti-reverse capability is a great feature. However, it is usually not wallet-friendly for the entry-level fly fisher (see Chapter Three on fighting fish and for more information on drag).

Our takeaway on purchasing a reel is to buy a decent fly reel ("decent" being the operative word), remembering it should be a semi-distant second in cost and importance to the fly rod itself.

Authors' note – A huge shout-out to Dave Waite from Ludlow, Kentucky for pointing out that the Kentucky Reel was invented by George W. Snyder (1780–1841) from Paris, Kentucky. This is the first fly reel invented in the United States. Snyder invented the reel in 1820 (without trademark or patent) and was quickly ripped off by every fishing company in the country.

Backing, Line, and Tippet

When you buy your first reel and get it set up, the folks at the store should explain backing, line, and tippet to you. Think of it in these basic terms: 1) backing is the stuff going onto the reel first, 2) fly line is the white, green, or orange stuff starting where the backing ends, and 3) leader/tippet is the six- to nine-foot tapered monofilament attached to the line at one end and a fly at the other. Tippet also refers to monofilament purchased on a small spool to be used to either add a second fly (dropper) from your primary fly or repair and extend tippet cut so far back the thicker portion of the leader taper is impeding the presentation. Most fly line is around 30-pound test weight, but the leader/tippet employed progressively reduces this strength down to anywhere from 2- to 15-pound test mono/fluorocarbon (7X to 0X) as you select. The X you choose will generally match up with your rod.

The size of the tippet depends on the size of the fish you are chasing. The larger the line is also a bigger tip-off to the fish. As we recommended, a good place to start is 5X. Personally, we have started using stronger lines for two reasons. First, we have found a bountiful honey hole with some very large fish. Unless you come fishing with us, we will not tell you where it is. Secondly, just like fish can see thicker tippet, so can we. Both of us are getting older, and a little thicker tippet makes it a bit easier to tie knots and untangle the line. The advent of fluorocarbon fishing lines (instead of standard monofilament) allows the use of greater strength leaders and tippets but also allows much greater invisibility to fish. This line strength adds to the ability of faster "catch and release" procedures which will be discussed later.

The Flies

The two basic categories are Wet Flies and Dry Flies – meaning some flies are tied to stay on top of the water, while others stay primarily below the surface. There are many subcategories of each but think of the depth of the water column you are fishing as the food chain for the fish. This will give you the first indication of what type of fly to use.

Without going in-depth about the whole life cycle and the metabolic history of aquatic insects during a natural fly hatch – you can find those videos online – flies are made to attract fish either underwater or the top of the water. The "hatch" is a body of water's "circle of life" to which you are attempting to match with flies to attract fish. Wet flies are imitating food in its various pre-surface fly stages. The dry fly is an imitation of the fly during the day; it is actually alive above the water's surface. Another type of wet fly is called a "streamer," which, when pulled (stripped) through/under the water resemble baitfish (or critters) your prey might chase.

- **Wet flies**: There are times when fish are shopping for groceries along the bottom, middle, or upper subsurface areas of the water column. This is where wet flies such as midges and nymphs may work well. When you develop

the skill to "strip" a fly through a subsurface water column, streamers become a fun option. Often, some adjustments in gear may be necessary to get your fly DOWN to the fish. For instance, sinking lines instead of floating, or changing leaders to longer monofilament, or sinking leaders to aid in the fly sinking to the area where fish are located. Saltwater fly fishing is a classic use of streamers and stripping the fly through the water to imitate baitfish.
- **Dry flies**: The fish are taking food from the surface or right near the surface. These dry fly offerings may include Adams, Caddis, Dun, and Wulff. Other flies can be considered "dry," such as terrestrials and poppers, as they stay on the surface.

One of your biggest challenges may be practicing your knots to quickly adjust to the latest hatch and reduce your downtime. We will get you practicing knots later in the book.

When choosing flies for a particular fishing trip, the choices are truly endless (and mind-boggling). Think of it this way, all the flies you see in the fishing store are potentially the fish's natural meal for the day. You must figure out which of those hundreds of choices will catch the eye (and hopefully the lip) of your prey.

G.S. Marryat said, **"The Essentials of a Good Fly-Hook: The temper of an angel and penetration of a prophet; fine enough to be invisible and strong enough to kill a bull in a ten-acre field."**

What works often changes from day to day and sometimes from hour to hour. This is why we always encourage fishers to drop by the closest fly store before they tie the first knot. If the person running the store tells you it is a nymph and midge river, you have your orders for the day. The store owner wants you to catch fish and come back for more fly advice and purchases.

Make sure you have enough of a variety to change out a couple of times. If a Copper John is getting no attention, switch to a Prince nymph and a Zebra midge. If the fish suddenly start rising to the top to feed, go to a dry fly. Trying to determine the next move of a fish is part of the fun.

We simply cannot finish this section without letting you know some of the flies in our boxes. These flies are generally geared for trout fishing. Other fresh water or saltwater fish may require additional trips to your local fly shop. The list of what is in our fly boxes is actually much longer than this, but these flies (along with the selections in "A Little Help From Our Friends") will get you started.

- **Dry Flies – (size 14 – 18 hooks, except Hoppers which may vary)**
    - Parachute Adams
    - Elk Hair Caddis
    - Blue Wing Olive and/or Pale Morning Sparkle Dun
    - Foam Hoppers – foam flies made to look like larger bugs floating on top of the water, eg crickets, spiders, etc. Also can work well as strike indicators.

- **Nymphs (size 14 – 18 hooks, except midges where smaller is better)**
    - Hare's Ear
    - Pheasant Tail
    - Prince Nymph
    - Copper John
    - Zebra Midge

- **Streamers (size 8 – 12 hook)**
    - Wooly Bugger
    - Muddy Minnow
    - Clouser Minnow

Net – Go to any sporting goods store, and you will find a plethora of nets. Personally, we like plastic netting over thread. Three or four minutes pulling a fly from a thread net is three or four minutes not fishing. When we are on a boat, we use a standard plastic net with a long aluminum neck. We also carry a cool net with a measuring tape built into the middle of it. With this net, there is no more guessing on the length of a netted fish. In any event, buy a net that is nonabrasive to the "slime layer" on the fish's skin.

Finally, when walking a rocky river or shore while fishing, we have both realized we are of the age when a fall could be catastrophe. When fishing on uneven ground, we use a walking stick or a net by Net/Staff that extends to become a walking stick.

"The fly angler who says they have never, ever fallen while wading, is either a pathological liar or has never been fly-fishing," author Jimmy Moore once declared. We have felt his pain. Literally.

Gear Packs/Vests – When you are standing out in a river and decide to change flies, you do not want to slog back to the shore to rummage through a tackle box. Having all your tools close at hand is important. There are several different types of vests and gear holders available. You will have to decide which one suits you best. Check out our links at the end of this book for further reference.

Waders – This is an expensive (and possibly unnecessary) purchase. If you are drift fishing from a boat, you do not need waders. If you are stream fishing or lake fishing from a shore, you do not need waders. Go crotch-deep in a cold river in the fall or winter, and waders suddenly become the best purchase you have ever put on your credit card. During the summer, we generally grit our teeth and wet wade. The cold lower body is offset by the heat of the upper body. But when the outside air temperature starts to get closer to the water temperature, you may need a good set of insulated waders to keep your legs and private parts from turning blue. When trying on waders, try to kneel and bend over as if placing a fish gently back into the water. If too tight, size up. Also, a word here about safety. Never wear your waders without your safety belt. If your waders fill with water, go straight for your whistle if you can. Avoid areas where your footing is potentially unstable and plan each step with safety in mind.

Multiple manufacturers are offering models and sizes to fit both men and women. Consider the amount of clothing you will most often wear underneath the waders, and it is a good rule of thumb to go larger than you think in order to have some freedom of movement. Most waders come with socks attached (not boots), so you can pick your own footwear. Two things: Again, buy boots 1-2 sizes bigger than your shoe size as the sock of the waders will be bulky. Also, we recommend rubber soles instead of felt soles. This allows for less transfer of invasive algae and critters from one fishing venue to another (yes, it is true, critters hitchhike). Use metal screw-in studs if you need more traction than rubber provides.

If you are staying shallow with your wading, hip boots are an excellent (and less expensive) option.

Polarized Sunglasses – A good pair of polarized sunglasses have two major functions. First, they take the glare off the water, allowing you to see the fish laugh at you as they turn up their noses at your latest underwater offering.

Secondly, glasses provide safety from an ill-cast fly. There are several techniques for removing a fly from a trout, and they basically apply to humans, as well. Moreover, the flipping and flopping of a trout as you slowly bring him in is a joyous sight. The flipping and flopping of your fishing buddy as you try to remove the same size fly from his or her face is not nearly as majestic.

Closed-toed Shoes – See safety warning issued with Polarized Sunglasses. Toes are just as susceptible to barbed hooks as foreheads. If you are not purchasing waders, a suitable option is a pair of hip boots or other waterproof hikers or ankle boots. At some point on any fishing trip every angler declares their final attempt at catching a fish. Cold feet should never be a reason for "last cast."

Sun Protection – Face, ears, arms, legs, and hands are very susceptible to the sun on a long hot day on the water. Please use 50+ SPF, long-sleeve shirts (neutral colors – no red or black), face stockings (a/k/a "Buffs®"), and a wide-brimmed hat. Skin cancer kills far too many anglers. Protect yourself and keep fishing for a lifetime.

Gloves – Consider the above suggestions for sun protection when it comes to your hands. However, keep in mind handling fish with gloves on can knock the protective slime off of them. Make sure you take a good look at our "catch and release" recommendations later in the book.

First Aid Kit – Keep the basics handy and, more importantly, know how to remove a hook from yourself or a fellow angler. Put monofilament around the bend of the hook, press down lightly on the eyelet, and then yank the looped line. It will hurt and bleed. If necessary, get medical attention as quickly as possible. The thought of removing a hook from someone's forehead is reason enough to switch to barbless hooks.

Life Vest – Do not drown. Wear a #%*! life vest. There is no excuse for not wearing one, so do not even attempt to rationalize this away. The auto-inflatable vests which activate when they hit the water will save your life. How do we know? We've both activated a vest while exiting a boat we suddenly found ourselves underneath.

Whistle – When fishing alone (or even with a partner), always carry a whistle. We hope you never have to alert someone you are in immediate need of assistance, but the whistle may save your life.

Miscellaneous – There are so many other things we have not mentioned: nippers, strike indicators, lanyards, forceps, slip shot, leader straightener, cheater glasses, extra strike indicators, insect repellant, needle nose pliers, goop, which cigars go best with different bodies of water....

The list is nearly endless, or should we say as long as your credit card limit. Therein lies some of the fun. Once you get into the sport, you will suddenly figure out what you really want to have on you when you go fishing.

Optional Adult Only Equipment

Cigars – Rick mentioned in his introduction his love of combining fishing with cigars. It is a habit we both enjoy. We do employ one rule - no smoking when teaching kids to fish. We all pass along enough bad habits to the next generation. Tobacco use should not be one of them. That being said, the question becomes what cigar pairs best with a day on the water. The answer is anything you like under ten dollars. There will come a point in your fishing life when you must choose between netting a hog and finishing a cigar. The choice being obvious, there is no worse feeling than tossing a freshly lit expensive stick (like an Opus X) into the water. Wade likes the house blend from Strauss Tobacco in Cincinnati, Ohio, while Rick enjoys the Excalibur from Hoya de Monterey (an old brand but available at most cigar shops.)

And while we enjoy our bourbon and scotch, please do not imbibe while on the water. Spirits are meant to sip while telling stories of the one that got away. Water and whiskey is a drink at the bar but a danger when fishing.

When putting out the money for your equipment, remember the words of Canadian author and chemist Orlando Aloysius Battista. **"Man can learn a lot from fishing – when the fish are biting, no problem in the world is big enough to be remembered."**

AND NOW … A LITTLE HELP FROM OUR FRIENDS

I used a fly rod from Walmart for the first year until I realized this is what I was going to do. I was not going to spend hundreds of dollars on something I was not sure I would like. But, of course, I spent money on my equipment after I found out I loved it.

—Sam Keating
Jacksonville, North Carolina

Tips on Starting
By Mark Forster © 2021
Program Lead, Project Healing Waters Fly Fishing
Lincoln, NE

Authors' note: This was modified from a text for the Project Healing Waters Fly Fishers class on selecting fly rod components.

Picking your first (or next) fly rod can be intimidating for the novice fly fisher. The number of options can seem overwhelming, and there's not much information available on how your choices will affect your casting or fishing experience. Fly shop sales staff can be hit and miss too; you could get a lot of great advice or just a lot of advice. Let's look at some of the variables you should consider when making a fly rod purchase (or selecting parts for a fly rod building kit if you're thinking about building your next fly rod).

Depending on where you buy your fly rod, you may have available options, including:
- Weight (Line Weight)
- Rod Length
- Number of pieces or sections
- Rod Action (or action speed)

Line Weight

Commonly, fly fishers start rod selection by deciding on line weight. A guideline chart below briefly covers fish species, hook size, tippet, leader, and line weights.

- Notice that heavy rods cover a larger spectrum of flies than light rods. While a heavy rod can throw a light fly well, that "advantage" can be deceiving as heavy rods lack the finesse, tippet protection, and ease of casting available from lighter weight rods/lines. If you are buying only one rod for multiple situations, fly rods on the heavier side of the range will certainly cast a larger variety of flies.
- On the opposite side of the spectrum, lighter rods (light end of the range) provide better feel, tippet protection, and ease of casting but run out of spine quickly when throwing overweight flies or heavier lines (like sinking lines). If you have not experienced a rod throwing too much line/flyweight yet, do not worry, you'll know it when it happens. The rod will feel like you're casting with a bungee cord.

One note of caution here: It is easy to think about a fish species you want to target, look up the recommended line weight, and buy that fly rod. Keep in mind that casting a much lighter weight rod than what you usually cast can result in a rod you may have difficulty casting (i.e., casting a 2wt rod requires a more precise casting technique than casting a 5wt rod). Conversely, casting a significantly heavier rod will take more effort than you may be used to exerting. Here, there is no substitute for experience with a wide range of fly rods. I remember wanting a 2wt rod for small stream trout until I cast one. I ended up building an 8-foot 4wt for small stream trout and being very happy with the compromise. There are 15 fly line weights but do not feel obligated to own a fly rod for each weight.

Rod Length

The average fly rod seems to hover between 8'6" and 9'6". Fly rod blanks run the gamut of length from really short (5-foot small stream) rods to very long (14-plus-foot Steelhead) rods. There are a few reasons to consider longer and shorter rods, many times specific to fish species or fishing environment. It is also important to consider your own casting ability before purchasing a rod for a specific environment.

If you fish where short casting is common (from a boat, in areas where you have a lot of surrounding or overhanging vegetation), or have difficulty with the amount of strength needed to cast a 9' rod, then a shorter fly rod may be the way to go. You lose casting distance but gain accuracy with a shorter rod, and they're typically easier to cast (from a physical strength aspect). It can be tempting to buy really short rods (under 7 feet) for smaller or younger fishers; however, those fly rods can be difficult to cast well, and you may be better off buying a lower line weight 8-foot rod to compensate for arm strength until the fisher grows into the heavier or longer fly rod. An example of this would be starting a fly fisher with an 8-foot 4wt fly rod until they can handle a 9-foot 6wt. The additional advantage here is that the 8-foot 4wt rod remains a solid choice for smaller streams and smaller fish species.

If you need to send flies a long way or you're stalking larger prey such as steelhead, then a longer fly rod may be appropriate. Longer rods let the caster gain distance but lose accuracy and ease of casting compared to shorter fly rods. Finally, remember that adding fly rod length will not compensate for a technically poor cast. It will only magnify casting problems.

Here too, casting as many fly rods as you can and taking notes is probably your best decision-making aid.

Number of Sections

I don't think most fly fishers give a second thought to the number of sections their fly rod has, perhaps because the most commercial fly rods are 9-foot, four-section rods. My take is this: Regardless of the type of ferrules a rod uses, every ferrule represents additional weight, loss of sensitivity, loss of flexibility, and potentially an increased chance of breakage. You may get to pick from two- or four-piece rods (or travel rods with more sections). My first choice is always a two-piece fly rod unless I need the compactness for air travel or backpacking. That being said, the minor performance differences would not discourage me from buying or building a four-piece rod. Additionally, if I were buying or building a rod for someone who didn't express a preference, I'd build the four-piece rod to ensure maximum flexibility.

Rod Action

The "action" or "action speed" of a rod is perhaps the least understood option when buying a fly rod. Generally, it is illustrated by showing the amount and location of rod flex under load. Faster action fly rods bend less than slower action rods, and that bend is biased toward the tip. Accurate information, but not particularly helpful in understanding the effect of rod action on a fly rod.

To understand action speed and its effect on casting, think about the fly rod as a spring. The action speed is how quickly it compresses and releases (or loads and unloads) with a given amount of energy. The best visual is probably imagining a slow and fast action rod held horizontal and then having identical weights dropped on each rod's line. The fast action rod would bend less, quit bending quicker, and rebound more quickly and with more energy (the weight would bounce higher) than the slow action rod. The quickness of rebounding is due less to the rod moving to absorb the energy from the falling weight (mostly at the top quarter to one-third of the rod length), and the additional energy returned to the line (bounce height) is due less to the fly rod moving during the rod's compression and release (the weight of the bent fly rod also must be accelerated). So that's an overly technical explanation that doesn't fully explain the effect of a fly rod's action during casting.

Examining how rod action affects casting, it is helpful to break down the cast into either a single forward or backward casting stroke.

- As you begin the backcast, the rod flexes toward the line until it is completely loaded. This is affected by the rod action, how much the line and fly weigh, how fast your rod tip accelerates, and how much of the rod mass bends during the casting stroke (rod mass that deflects during casting is "sprung" weight and uses energy when the rod unloads). When the rod is fully loaded, or reaches equilibrium, the line speed equals the tip speed. At the end of the casting stroke, the line speed is increased by the rod unloading. Assuming that either rod is fully loaded, the fast action rod will add more speed to the line and throw a smaller loop (loop size is determined by the arc of the rod tip, a stiffer/faster rod deflects less and therefore throws a smaller loop) than the slower action rod. Pausing here for a moment, consider that with short lines or a slower casting motion, the fast action rod will be difficult to load fully. This is the first area where the slower action favors beginning casters or fishers with a slower casting stroke.

- At the end of the casting stroke, the line will unfurl to the rear of the fisher. When the line is fully unfurled, it will pull on the rod tip and begin to load it (you feel this in the grip of the rod) with excess energy left in the line from the backcast. Starting the forward casting stroke should occur sometime between when the line is unfurled and when the tip begins to move forward or unload. Once the tip begins to unload, the line will develop slack or start to loop which will take energy away from the forward cast. This is the second area where a slower action rod favors beginning casters.

- Note: Using a line 1 weight over the rod listed weight can slow action by increasing the relative load applied to the rod during casting. This extra load creates more deflection in the rod tip during casting which increases the amount of time available to reverse the cast. The effect of overlining a rod is limited as it is still possible to overpower the spine of the rod with heavy flies and long casting loops.

- Other considerations for rod action (assuming your casting is technically sound with different action speeds):
- Aggressive, head-shaking, and tough-mouthed fish (muskie, pike, gar, bass) may be easier to land with a faster action rod. The rod's stiff spine and quick recovery from loads allow for strong hooksets and keep pressure on the fish while it's fighting.
- Fishing heavy lures, sinking lines, long casts, and windy conditions also favor the fast action rod. Fast action rods can absorb larger loads, are more efficient in casting, and throw smaller loops (beneficial in windy conditions).
- Soft-mouthed fish (like crappie), light tippet, and lightweight hook situations favor the softer spine of a slow action rod.
- Casting with less backcast may also be improved with a slower action rod since the rod is easier to load under those circumstances (shorter casting loops).

As a beginning fly fisher, it's easy to find yourself with an expensive fly rod that offers features you don't need or can't use. Armed with a bit of knowledge, you can easily find an inexpensive fly rod that helps you learn proper technique and increases your chances of success.

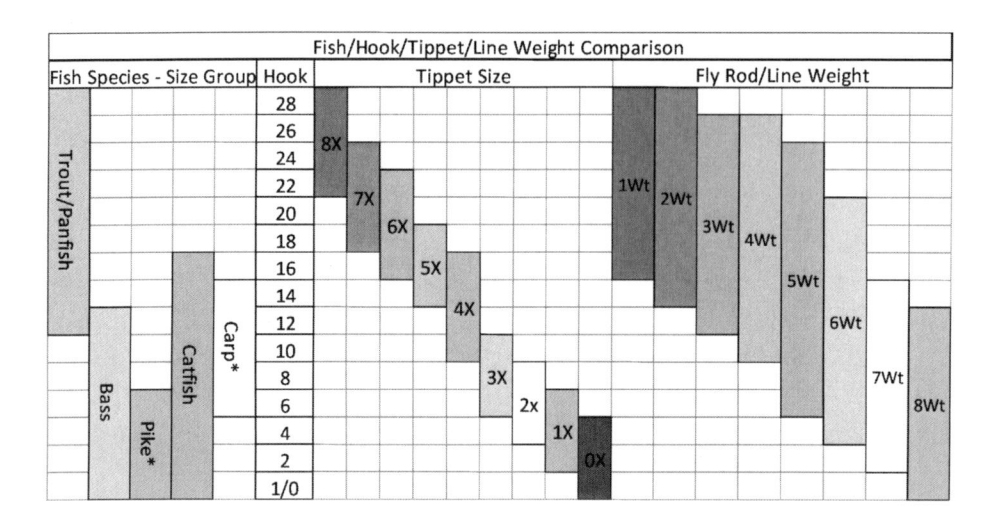

Authors' note – Consider contributing to Project Healing Waters Fly Fishing or assisting with their programs.

Project Healing Waters Fly Fishing is dedicated to the physical and emotional rehabilitation of disabled active military service personnel and disabled veterans through fly fishing and associated activities, including education and outings. The program is not a one-and-done experience but rather an ongoing effort that relies on local fishing clubs to conduct programs. Check out the links in Chapter Nine to find a participating club near you.

My experience comes from fishing a small burn, which is a Scottish word for a stream, that runs into the River Clyde. It is a small water that runs off the Moors south of Glasgow with small, hard-fighting fish that can be shy to the hook. Due to large waterfalls, salmon and sea trout cannot get upstream. The fish here are free-rising and fun to catch on light fly tackle 2 line and 6ft rod. The flies I use are small black spiders and green Wells pheasant tail nymphs. I also use a 1 pound breaking strain line. The person who taught me about fishing was an old man with old gear, and what he taught me could not be bought.

—Barry Carruth
Glasgow, Scotland

---

1. If you are starting as a rank beginner, don't buy high-end equipment to start. A good beginner outfit is all you need to see if you like fly fishing. That way, if you don't end up liking it, you're out maybe a couple of hundred bucks.

You will never get anything near what you paid for the equipment if you sell it.
2. Pretty loops are great, and learning to cast well is important, but fish don't care how beautiful or technically accurate your cast is. They only care about what you put in front of them. Get the fly in/on the water.
3. If it's raining, you can still fish. The fish are already wet.

—Greg Lumpkin
Chesterfield, Virginia

What Is Tenkara Fishing?
By Spencer Durrant ©

When tenkara fishing came onto the scene, I dismissed it as a fad. Years back, when I had a regular column in a local newspaper in Utah, I even wrote that I didn't think tenkara would last long.

Boy, was I wrong.

Tenkara is here to stay, it seems. And just for the record, I don't have any problem with tenkara fishing. It's a great way to introduce people to the idea of fly fishing, even if it only roughly resembles fly fishing.

So, what exactly is tenkara fishing, and how does it differ from conventional fly fishing? Let's take a look.

An Overview of Tenkara Fishing

Tenkara fishing is a style that originated in Japan, specifically in the mountainous region of that gorgeous country. There, fishers perfected the art of using long, thin rods with level lines and a single fly to entice fish to bite. There is no reel, and the line you use is a fixed length.

According to DiscoverTenkara.com, it wasn't until the 1980s that tenkara got its formal name and was established as a legitimate form of fly fishing in Japan. Since then, tenkara has found a good foothold here in the United States, especially on the smaller streams in the Rockies.

Tenkara rods tend to be much longer than a conventional fly rod and even put some Euro nymphing sticks to shame. A rod of 11 to 14 feet in length is the most common in the tenkara world.

At first glance, you might think such a rod is too heavy and therefore unwieldy. However, tenkara rods don't use line guides, reel seats, or other hardware. That reduces the weight of the rod overall and allows for a lighter, more responsive rod that also gives you the length to lift almost all of your line off the water when presenting flies. For small streams, that's a dream for those of us who know that big trout are just waiting for a perfect dead-drift in a back eddy or tucked in an undercut.

Most tenkara rods are telescoping, too, which provides an ultralight fishing solution for the high country. As I spend more time in the high country, I run into more and more backpackers who brought a tenkara rod along for fun and caught enough fish for dinner.

Finally, you have the fly used in tenkara. The traditional name in Japanese is "kebari" which translates to "feathered hook." How apropos.

The kebari is essentially an inverted soft-hackle fly. Instead of wrapping the hackle around a fly hook with the fine hackle points pointing towards the back of the hook, a kebari has the hackle fibers pointing forward towards the hook eye.

And that's all you need for tenkara fishing. There are no reels, no fly floatant, no extra line sizes – just a few flies, a rod, and a level line. It's a minimalist's dream and isn't as limiting as it might appear at first glance.

The Kebari

When I tried tenkara for the first time, the kebari interested me the most. I've always been a fly tier – my grandfather tied commercially for 27 years, and he's responsible for both my dad and I tying for nearly as long – and the kebari looks like a beginner's mistake.

It's not – the kebari is an ingenious idea.

First off, a kebari doesn't attempt to imitate specific flies. This is largely due to where tenkara itself originated. The high mountain streams of Japan are relatively similar, ecologically speaking, to the high country of the Rockies here in America. The streams are gorgeous, but most high mountain waters aren't as fertile as tailwaters or lower-elevation fisheries. Here, the hatches are smaller, and fish don't have the luxury of being picky when eating food.

The Japanese anglers who developed tenkara recognized this phenomenon hundreds of years ago. Since they were fishing for survival and not just for fun, they needed to develop the easiest way to put fish in the net. They found that a wet fly, presented correctly, caught plenty of fish. So, why bother with boxes of dozens of patterns when you find one that works well?

The kebari is designed to imitate wet flies, and the reverse hackle creates more friction against the water's surface. This reduces line sag and increases the chances of hooking fish – an important feature when fishing for survival.

Now, you're not limited to just a kebari. Many tenkara anglers use dry flies, nymphs, and other wet flies that you'd find in any fly shop in the United States. The fly you choose depends mainly on the type of water you're fishing and how you want to fish.

The Line

Tenkara uses two primary types of lines – level and furled. Level lines are exactly what they sound like – stretches of line with no taper. Furled lines are similar to furled fly-fishing leaders. They have a taper but are much longer than a simple furled leader.

Furled lines generally give you a better presentation if you're fishing dry flies.

Level lines are so simple that you can virtually make your own and cut them down to any size needed.

Line length matters because, like in traditional fly fishing, you attach a tippet section to the tenkara line. Jason Klass, a former fly fishing guide and casting instructor who writes the blog Tenkara Talk, explains that he usually fishes a 10- to 11-foot line, with another 2- to 3-foot section of tippet.

Your line length depends entirely on the style of fishing you're doing. For example, if you're fishing nymphs, you'll likely want to use a shorter line – maybe 5 feet – with double that length of tippet. This enables you to achieve a nearly perfect copy of a Euro nymphing presentation.

If you're fishing dry flies or in other situations where you need more stealth, then a longer line is what you'll want to use.

The Cast

Casting a tenkara rod proved more difficult than I expected because I kept trying to cast it as I would a traditional fly rod. However, tenkara casting is effortless – once you get the hang of it I managed to figure it out after 15 minutes of frustration, and if I can figure it out, anyone can.

Wrapping Up

Tenkara fishing is an interesting method of fly fishing, and it's a good option for those looking for an ultralight fishing method. It's certainly here to stay, and if you ever have the chance to try tenkara fishing, I highly recommend it. It'll reinforce the good habits you need to achieve great presentations in traditional fly fishing if nothing else.

Spencer Durrant is a fishing writer, guide, and bamboo rod builder from Utah. He's the Owner/Lead Guide at The Utah Fly Fishing Company, the News Editor for MidCurrent, and a columnist for Hatch Magazine. Connect with him on Instagram/Twitter, @Spencer_Durrant.

Lucy and Ethel Go Fishing

As a woman growing up with a father who did not fish, I often surprise many people when they learn that I am part of an annual women's only fishing tournament in the Northwoods of Wisconsin. My fishing excursions are often comical but also very successful. There are plenty of tangled lines; snagged logs, rocks, and trees; and fishing poles and sunglasses lost to the bottom of Pier Lake. But I also have a few trophy fish on the wall, too.

One year my friend Amy and I (my name is also Amy) decided to pre-fish to locate the best fishing spots in advance of the tournament. We could not get the boat started, so we borrowed the neighbor boy's paddle boat. We paddled to the spot, caught fish, and realized we had nothing to put them in. We paddled back, got a high-tech live well (5-gallon bucket), and paddled back to our spot. We caught a few more fish and realized we had committed a sin in Wisconsin: We forgot our cocktails. Although I have never actually seen it in print, I have always assumed it is a state law not to fish without alcohol. We paddled back to get cocktails. Life is good. We paddled back to our spot, caught some more fish, and realized we did not have sunscreen. Paddle, paddle, paddle! We could not have had more laughs, and we caught enough fish for supper that night and even more for future meals.

Since we had overfished our spot the day before, there were no fish on the day of the tournament. Lesson learned. The bad news is we lost out on the potential prize of $40, which is about what a round of drinks costs at the sponsoring bar. Nonetheless, we had a day of pretending we were Lucy Ricardo and her crazy friend Ethel.

—Amy Bliss
Cottage Grove, Wisconsin
Authors' note – Well, we did say a good attitude is the most essential piece of equipment a fly fisher can pack.

It is Not the Age of the Fly that Counts ...

On Jan. 17, 2015, I went fishing for trout in the Cumberland River just below the dam for Lake Cumberland. It was very cold as our fishing rod guides were freezing up, and I had to clean the ice off of them repeatedly. That's dedication to fishing. The Army Corps of Engineers was not generating hydropower, so there was little current flowing. I was in my friend's 21-foot Deep V Tracker Boat with him, his son, and his son-in-law from Somerset, Kentucky. He had a lot of experience fishing the river as well as the lake. I have great respect for the Cumberland River in a boat, as it can be dangerous. I was using a homemade rod made by a friend of mine with whom I fish for walleye on Lake Erie. It is of good quality material, medium action with a fast tip. The reel was a Shimano Sustain 1000. I had it rigged with a minnow on an Aberdeen number 2 hook with a number 2 split shot about 16 inches above the hook. My friend had the boat positioned against the dam—really! I even reached out and touched the concrete dam. Wow! We were just casting out and slowly reeling it back. We caught several trout that day. Some were caught using a 2-inch black and silver A.C. Shiners crankbait. My trout was caught while using the live minnow method. I didn't realize what I had until I got it in the boat and my friend said it was the biggest one ever caught in his boat. The trout was 22 inches long. I did not weigh the fish. It now resides mounted on my living room wall.

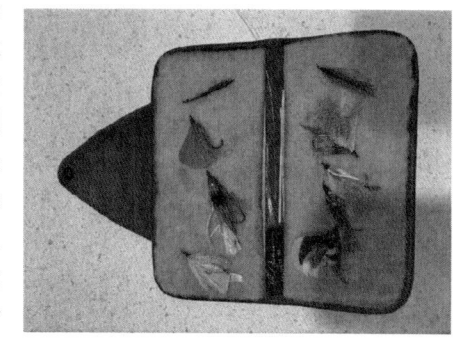

The previous day I had gone striper fishing by myself in Beaver Creek, and I caught two keeper stripers, 26 and 28 inches. I also caught a nice 21-inch smallmouth. Those were a great couple of days of fishing.

The flies were bought in an antique store in Eastern Indiana. They were from L.L. Bean. I have been to the home store in Maine, so I thought it was cool to find them.

—Tim Smith
California, Kentucky

The Annual Tune-Up

Doing an annual gear checkup, cleaning, oiling, etc., is especially important. Here are a few tips that I follow. This is not something I found on a website.

1. I use WD40 on all my moving parts.
2. Cork grips –Mr. Clean Magic Eraser is the best way to get them looking like new. After the grip is clean and thoroughly dry, apply several light coats of Tung Oil to the cork. Let each coat dry for a day or so. I usually apply three coats, hand-rubbed. Remember, cork is wood!
3. If you want to protect your expensive reels, use Boeing's Boeshield! Expensive and sometimes difficult to find, this is the Cadillac of metal protectants. I have used it since the 1990s. I originally discovered it through my woodworking endeavors. I used it on my scroll saw, table saw, and drill press. It's the No. 1 product pros recommend. I use it for the bodies of my reels; after they have been allowed to dry thoroughly, I use the WD-40 on the springs, crank, etc.

Most of my gear is considered either "old, antique, or collectible." (Sort of like me!)

The Tung Oil will make your grips last a lifetime, and they will stay a lot cleaner. Tung Oil comes in many natural wood colors (walnut, cherry, maple, etc.) and can be applied to change the color of your grips.. Have fun but remember, when you change the color, it is permanent.

—Joel Stansbury
Florence, Kentucky
Authors' Note – Clean your line regularly, too.

The best tip I have is if you can't walk on water, wear hip boots.

—Geert Sandman
Almelo, Netherlands

Filling that First Fly Box

If you take the opportunity to utilize a guide in your early fly-fishing adventures, he or she will have several things that will be starkly different from your equipment. One of those, no doubt, will be his wide array of totally packed fly boxes. Relax, you do not need to match those boxes just yet.

So, where do you start? Most fly fishers start with two boxes—one for dry flies and one for nymphs. Later, you will want more boxes and break them out in one of several ways. There really is no wrong, just what works best for you. Some will do boxes by the most popular bug species on their nearby rivers. Others by size, stream, fly types, etc. It's whatever allows you to find what you are looking for the easiest and fastest.

You will also notice there are dozens of fly box configurations, types, materials, etc. Again, they all work. It's a matter of preference, which you will develop over time. So choose two simple, inexpensive fly boxes that you like for how they function. I suggest you start by using one for dry flies (flies that float and imitate mature insects or terrestrials). The other you will use for nymphs and streamers, flies that sink and are fished under the surface.

Once you've got your boxes, start by going to your favorite local fly shop and asking the workers what the top five dry flies and nymphs are for your area. If there are various waters in your area, be specific and ask for the top five flies on the rivers or lakes you plan to fish regularly. While there are old traditional "go-to" flies that work almost anywhere, your area will have specific flies that simulate best the bugs common to your rivers and lakes. Buy a dozen of each of these, and then do the same for the top five nymphs for your area. Brace yourself: All of this will fit in a few small plastic cups and cost you more than you will be expecting.

About this time, you'll be tempted to start buying flies from unknown online vendors at much cheaper prices than the well-known brands most fly shops carry. While good deals can be had, many of these flies are tied with low-quality hooks and materials. Plus, the high-volume tying required to make flies at these low costs usually produces an inferior product. I've spent countless frustrating minutes trying to clear the head cement out of the eye of the hook or trying to thread leader into a fly where the material is tied too far forward. Buyer beware.

Put these lures in your box. Yes, your boxes will still look empty. They will fill with time. What you will want to finish out your boxes with are the flies you actually use most. You'll do this by growing your box every time you fish.

As you head to the water, always stop at the local fly shop and tell them where you are going and when you will be there. Ask them what they would suggest. Flies are the same in almost every fly shop and usually cost within a few cents of each other. Fly shops don't earn your business by having "better flies." They will keep you coming back by giving you "better information on flies." While you have only a few boxes and don't know the names of all the flies well, take your fly boxes in the shop with you. Check your box, so you don't buy flies you already have. Then buy a half dozen (or maybe a dozen if the budget allows) of the 3 or 4 flies you don't have that the shop recommends.

If you do this each time you go to the water, you will soon have a box full of flies that work for you in your area. You'll find you have favorites other fishers don't like. And you will find flies your friends love but just don't work for you. Soon you'll have your go-to flies that you will want to keep in various sizes, colors, etc. And yes, before long, you'll want more boxes for more flies.

Treat these boxes like a prized possession. They are a unique reflection of you as a fly fisher. But eventually, that day will come when you realize just how valuable they really are. As you rush to change flies, you will return that box to a pocket, failing to zip it up. Later, when you realize it's fallen out somewhere during the day, you will initially think about the loss of the box, then, a few seconds later, the crushing fact that the $30 box was nothing compared to the hundreds of dollars of flies it contained.

Then you'll get to start the process all over again.

—Jim Young
Woodland Park, Colorado

3

How The Heck Do I Use All This Stuff?

"When trout fishing, one must be a stickler for proper form. Use nothing but #4 blasting caps, or a hand grenade, if handy, or at a pool well-lined with stone, one blast from a .44 magnum will bring a few stunned brookies quietly to the surface."

—Edward Abbey

This chapter is our attempt at giving you the information you need to get out on the water and start using the basic gear now sitting in the trunk of your car.

Before starting this chapter, we would like to point out you need more than words to learn what we are discussing. Check with your local fish and wildlife departments for clubs offering instruction. For those wanting to improve their skills at fly fishing, we strongly recommend signing up for Fly Fishing 101 and 102 at your nearest Orvis store.

When taking these classes, remember the words of Charles Orvis himself:

"Unless one can enjoy himself fishing with the fly, even when his efforts are unrewarded, he loses much real pleasure. More than half the intense enjoyment of fly fishing is derived from the beautiful surroundings, the

satisfaction felt from being in the open air, the new lease of life secured thereby, and the many, many pleasant recollections of all one has seen, heard, and done."

We can tell you how to tie knots, but the instructors at these schools put lines and flies in your hands. Many of these classes are free and give hands-on instruction. They also offer beginners' fly-tying classes. Both of us are graduates of Orvis' daylong schools, as well. But these lessons are merely a means to the end of enjoying a life in context with nature.

Knots

Losing a big fish to a poorly tied knot will break your heart quicker than a high school prom date. New fishers (fly or cast) are always getting their waders in a bunch about knots. And the frustration is for good reason. There are so many knots fellow fishers will recommend, it will be hard to remember them all.

Wade spent 30-plus years on the job as a firefighter. In his line (pun intended) of work, the importance of learning to tie a good knot is an understatement. For him, knots could be life-saving. Knots were his friend to get him out of tight spots. Knots were practiced so they could be tied with poor visibility, under duress, and as if your life depended upon the knot NOT coming undone.

Now, tying fishing knots is not quite as dramatic. Still, if you lose a big fish because your knot comes undone or breaks because it was not neatly tied, there will most likely be a minimum of gnashing of teeth and possible references to the fish being some son of a satanic figure.

Authors' note – Remember if a friend is filming you catching a fish, your children may well hear the language you use when losing a fish.

Before we get you started practicing some knots to use, you can thank us later for these fundamental knot-tying tips.

Tips for Practicing Knot Tying

First, practice tying your knots with a piece of ¼-inch (or larger) rope. Find a piece of soft cotton (not poly) rope about six feet long. So you can "see" the assembly of the knot as it comes together, practice these knots with the bigger rope. After you practice with the big stuff, it will become readily apparent that tying knots with 5x tippet is a whole new game of fine motor skills, dexterity issues, line stiffness/memory challenges, and the visual feat of tying a neat and strong knot with a fine thread of line. But the movement of the line remains the same.

As some knots will weaken the line's strength more than others, choose your knot wisely. Friction and binding of the knot will heat up during tying until eventually the line cuts through itself, and there goes your fish – probably with a fly or lure stuck in its mouth. In late December of 2020, Wade's son caught the same fish Wade had hooked and lost a half-hour earlier. The 18-inch rainbow still had Wade's prince nymph and the midge dropper embedded in his lip. Father and son quickly relieved him of both hooks and sent him back into the creek in fine shape. Wade blamed it on his worsening eyesight and used the flies again – this time with better knots. The day pretty much summed up the Karma of the year 2020 for everyone.

Secondly, tie a neat knot. This means keep the wraps and tucks orderly as they are part of the knot's shape. This is called "dressing" the knot. One of Wade's old Scoutmaster friends, Don Berger, was a Merchant Marine during World War II. Needless to say, Don could tie more kinds of knots than either of us could list, but one of his sticking points during instruction to his Scouts was to "dress your knot." As we believe any self-respecting trout can see and reject messy rigs, practice dressing your knots.

Third, slobber on the knot before you cinch it tight (not the rope you used for practice). Spit is always handy and absorbs much of the heat generated by a monofilament line being cinched tight into a knot. The wetter, the better, and do not worry – the lake, creek, or river will wash off your slobber-scent clean as soon as it hits the water. Who knows, maybe the next fish will like the pastrami sandwich you just finished.

Three Recommended Knots (and one not recommended)

There are more knots than we could ever possibly teach a fisher to tie. Many are just as good as what we list here, but keep in mind you are working with fine leaders and tippet, so some knots may not work on a #18 midge hook eye. Take the time to watch website video links for knot tying. And if you are a visual learner like us, animation videos are hard to beat. A fisher usually has their "go-to" knot they become comfortable tying. We are no different and recommend learning at least the following three:

The Improved Clinch Knot:

The clinch knot is a basic knot used forever by most fishers. If you are a spin fisher, it is used for attaching hooks or lures to your line. If you are a fly fisher, the same concept of the clinch knot applies to tying flies to your tippet. It is easy to tie, and we suggest you learn this knot first as it is a strong (and simple) knot.

Put the fly in your non-dominant hand between your thumb and forefinger. Thread through the eyelet of the hook and pull about 3-4 inches of line through the eyelet, doubling back the free end against the line leading to the rod. As you begin to learn knot tying, you may need to pull more line through the eyelet. Twist the free end of the line five times around the line leading from the fly to the rod. Pass the free end of the line through the small loop formed just above the hook's eye, then through the big loop you just created. Spit and pull tight. Cut the excess of the free line from the fly. When you mess up and cut the other side, do not get upset. We have all done it. Thread the hook and start over.

The Surgeon's Knot:

This surgeon's knot is used to connect lines, usually tippet or leaders. It is strong enough for most stream fishing. Other knots offer more strength, but unfortunately, those knots are more difficult to tie. Save learning those more complicated knots until you have mastered the surgeon's knot. Tying a surgeon's knot will save you not only time but also money. When the finer tag end of a leader starts getting short from too many clips, the surgeon's knot allows you to add more tippet and quickly get back on the water. When you start fly fishing, you will tend to buy a lot of new pre-packaged leaders, beginning thick and tapering to a fine tippet. Once you learn how to add a couple of feet of tippet from a spool, you will have wondered why you (and your wallet) waited so long to add this knot to your repertoire. Double and triple surgeon's knots are easily tied.

To tie tippet to your line using the surgeon's knot, place each line side-by-side with a couple of inches overlapping each other. As with the clinch knot, you may need to use more line and tippet when learning. Bend the overlapped line and tippet to form a single loop. Here is the hardest part – pass both the tag end and the leader through the loop. For a stronger knot, repeat this pass-through two or three times. More spit and pull tight. If everything slips through on the pull, take a deep breath and try again. As we tend to reiterate, if you are visual learners like we are, watch any internet video on the surgeon's knot to learn how to tie it quickly.

The Mono Loop Knot:

The mono loop knot lets the fly or lure "hang free." As it has a loop through the eye of the hook, not cinched tight, it allows the fly to act more lifelike. Many swear by this advantage as a better fly or lure presentation knot, and it is easy to tie! It's designed for smaller line sizes like the stuff trout fishers tend to go after but can also be used on much higher pound-test leaders and tippets, as well.

The mono loop knot is engineered to be nonslip and leave a space between the knot and the fly. Start about 8 to 10 inches from the tag end of the line (a little more or a little less, depending on the length of the loop you desire). Make an overhand knot in the line and then thread the tippet through the fly hook eye and then back up through the loop of the overhand knot. Wrap the end of the tippet around the line five times. Now thread the end of the tippet back through the overhand knot. The tippet will enter and exit the same end. Apply even more spit and pull to set the knot.

Experience Tip: As you cast, snag on rocks and logs, drag your line over obstacles, catch fish, or otherwise accumulate stress on your line, knots and tippets and leaders may weaken and eventually cause aggravation. It is not unusual for knots to fail (if not dressed properly) or break from stress and cutting through themselves. After fishing hard, slow down, take a break, and re-tie your rig with fresh leader/tippet and good knots. The next fish might be your personal best. You do not want it swimming away from you with your favorite fly hooked in its lip while you look at a squiggly inch of unraveled knot tippet.

The Wind Knot:

There is a fourth knot (for which there is no instruction) that every fly fisher ties regularly. It is called a wind knot, which is a nice way of saying you tangled up your line with a bad cast or retrieve. Do not get upset. We all get tangles.

The wind knot results mainly from one of two mistakes. Either you start the forward motion of your cast too soon, or you miss a fish, and on the retrieve, you do not complete the backward motion of the ensuing cast.

When you get a wind knot, you have three options. The first option is untangling the knot. The first time you do so, you will be amazed at how many tangles can result from the flight of a single small fly. If you are fishing with wet flies, it sometimes is easier to untangle after removing the strike indicator.

We usually work about five minutes on a tangle before moving on to the second option: using your nippers to cut the line as far down as possible. This method often leaves you short on the tippet end. Refer to our instructions on the surgeon's knot to add more tippet.

The third option is to simply cut the line off at the loop and start over with a new line.

Every time you cut line, put it in your pocket or sling bag. Do not put cut-up line in the water. It is bad for the fish and incredibly frustrating for some fisher who will eventually land your discarded line instead of a fish. They may have never met you, but they will not say nice things about you whatsoever.

The best way to avoid wind knots is to practice casting. Many people practice in their backyard with a homemade Velcro fly and a block of wood covered in Velcro adhesive fabric. Our favorite place to practice casting is on still water. First, you can generally find differing wind directions to develop variations on casting with and against the wind. Secondly, practice casting on still water occasionally comes with the added benefit of actually catching a fish.

And speaking of casting...

Fundamental Casting Technique and Types

Do not fret ... here is where it gets fun. You have rigged your rod and reel with backing/line/leader/tippet. At this point, we are assuming you can get by with tying a few of your own knots and have a few boxes of flies and a strike indicator (bobbercator) or two. Now it is time to learn some fundamentals about various casting methods, and we are going to start easy to build confidence.

Our simple intent regarding the cast is to reduce the potential frustration you may encounter on the road to becoming relatively effective casting a fly rod. There are several types of casts and a long list of video instruction on the internet offering guidance on proper mechanics. By all means, knock yourself out and practice these basics with the same vigor you would use to pursue a large fish. Our casting advice is designed to help avoid some common pitfalls and offer solid basic casting concepts and options which can help you succeed in any fishing environment. Practice and good old experience on the water will sharpen the skills from there.

- One of the greatest fly rod casters won 21 national championships for distance and accuracy during the 1940s and early 1950s. This fly fisher's longest cast was an incredible 161 feet. She accomplished this when there was not even a women's category in competitive fly casting. Joan Salvato Wulff still runs the Wulff School of Fly Fishing today, and her internet videos are easily accessible. They will expand your casting knowledge and reinforce the need for technique over power.

For the beginner, we offer some rules for fundamental casting techniques. This is the part of the instruction manual which usually states before you go out on your own: "**READ ME FIRST.**"

Fundamental Casting Technique –
Rule No. 1: Form over Power

As is true with many physical activities, work on proper execution rather than power. Greater accuracy, distance, and proper presentation of your bait/lure/fly are derived from effective body mechanics. Once you get the form mastered, you can always add more energy to your execution. Early in your fishing, using too much power to compensate for poor execution usually results in confounding the desired outcome (and a lot of wind knots). If you are a guy, throttle back on the testosterone, and if you are a female, you have an advantage here from the start.

As an example, we have made many references to the similarities between golf and fishing. When we were stronger and younger men on the links, we both remember being soundly thrashed by 80-year-olds in our foursome. We tried to show off and hit the ball as hard and as far as we could while the older folks just made good contact with the club and put the ball right down the middle of the fairway. Usually, they were on the green and in the cup while we were still looking for our ball in the woods. Being effective at fly fishing is quite similar to the octogenarian golfer. It is a marathon, not a sprint. Technique is your friend.

Fundamental Casting Technique –
Rule No. 2: "Loading the Rod"

We've all probably heard the old saying, "timing is everything." We have mentioned taking classes to assist with ingraining good muscle memory and habits. We hope you took advantage of what instruction (classroom or virtual) was available, but practicing those classroom concepts should be continued to develop accuracy and consistency (See Rule No. 1). Poor loading technique is also one of the leading causes of the dreaded wind knot.

When loading the rod, compare the difference between casting with spinning tackle and fly tackle. When we cast with spin fishing equipment, the lure goes out and the line follows. When we cast with fly fishing equipment, the line goes out and the lure/fly follows. It was Joan Wulff who said when fly fishing, "**we must learn to cast backwards.**"

Remember when we told you the rod was your most important piece of equipment? To understand "loading the rod," take a moment to remember that with a fly rod, there is not (usually) enough weight on the end of the line to pull the line out toward the desired target of the cast. You will quickly learn the weight driving the cast comes from the length of the line/leader/tippet/fly beyond the tip of the rod. This is what "loads" or bends the rod at the end of each forward and backward stroke of the rod. THEREFORE, keep in mind if you start going forward or backward with your casts too quickly (before the length of line turns over completely), you will hear the fly end of the line snap or pop like a bullwhip.

If you have a short length of line out, it will allow for faster repetition of forward and backward. By the way, the old rule of thumb is to stay between 10 o'clock and 2 o'clock when casting a fly rod. While this was once widely accepted, it may not be as "hard and fast" a rule as we once embraced. Longer lengths of line require greater patience (wait for it…) and timing practice to achieve. Start practicing and fishing with short line lengths and add more line/distance as you grow mentally and physically confident in your skills. If you find yourself taking the rod back too far on the load, try tucking the butt-end of the rod into your shirt cuff as a reminder to stop at 2 o'clock.

Fundamental Casting Technique –
Rule No. 3: Release to a Target

Slinging a fly in the general direction of water is not casting at a target. Standing on the banks and casting so close to a rising trout that a slight shift in the wind might cause the fly to give him a concussion is casting to a target. This should be the goal when practicing your cast. Adjust your target to an upstream landing target so that your fly drifts naturally into the fish's feeding lane resulting in a better chance at a strike.

If you remove the chronological age of each of us from the challenge of learning to fly cast, you may want to use the same exercises many used as kids. As young children, many of us were tasked with practicing casting a weight into a 5-gallon bucket.

We never knew if it was practice, a game, or just a way to keep us busy for a while. In any event, it was a great way to understand the mechanics of timing – "loading the rod" with energy to send the lure in the right direction. The same practice can be used for a fly rod, but in this case, the line itself becomes the weight sending the fly towards the target. Get out in an open area where you can practice casting and make a 3-foot target circle with a piece of rope. Remember, when your line is already on the water, additional drag will occur and can be used in your favor to "load the rod."

Of the several types of casts you can utilize on any fishing trip, the type of cast you choose depends on how much room you have to work and how much you actually need. The following fundamental deals with three basic types of casts to get you started. The overall goal in casting is to gently land your flies in the desired location and get as lengthy of a drag-free, natural looking drift as possible for the fly.

Fundamental Casting Technique –
Rule No. 4: The "Standard" Fly Cast

The "Standard" Fly Cast is the classic backward and forward casting method commonly seen in all the pictures and movies of fly fishers. It is used to gain accuracy (direction and distance) to the target. As a general rule, try to keep the back-and-forth stuff to an absolute minimum. Do it just enough to get the job done. The more movement above the water alarms fish and can often reduce your chances of a strike opportunity. This rod loading is critical to understanding proper future use of the rod. Forward casting too soon or using too much wrist are early bad habits to avoid. See Technique Rule No. 2.

Fundamental Casting Technique – Rule No. 5: Mending Line Casts

"Mending" the line means to lift the slack in the line upstream on moving water, so the fly doesn't "drag" quickly through the water. In other words, when you land your cast on moving water, consider immediately correcting the slack part of the line upstream to lengthen the time of your drag-free drift. Mending line can be utilized while wading or boat drifting as needed. Mending can also be your friend. If casting into a tight spot where a fly might get hung up, the mend will catch the faster water and drag the fly out quicker than the slow water around the rocks. This will make for a better presentation and the safe recovery of an otherwise hung-up fly.

Fundamental Casting Technique – Rule No. 6: Roll Casts

A roll cast is a very handy cast to employ when you are in a "tight spot." The cast is a "rolling loop" in your fly line, allowing you to send your fly out to the desired location without backcasting into the trees or bushes behind you, just rolling the wrist to flip the line. When fishing in boats or shorelines where your backcast will get tangled or possibly hit/hook somebody, learning to be accurate with the roll cast is a great advantage.

Fundamental Casting Technique – Rule No. 7: Casting for the Illusive Tree Bass

You WILL get hung up in a tree. No one knows the name of the fly fisher who caught the largest tree. Knowing what to do when this happens will enable you to eventually laugh at the size of the tree you landed.

Do not yank the rod sideways to free the fly. It embeds the fly deeper into the wood, and your rod is not made to withstand the strain. Do point the rod directly at the miscast fly and pull on the line. Half the time, it will snap the tippet (better than snapping the line). The other times it will free the fly. In both instances, the line will come back at you very quickly. Be safe when trying to free flies from trees. It is better to catch a "tree bass" than grabbing our book to remember our instructions on removing a hook from under your skin.

Time to Practice Casting

At this point, consider going outside and practicing. The following practice points we offer make no assumptions on what you may already know or not know about fly rod casting. Like we have said repeatedly, there are tons of fly-casting videos on the internet, and you can watch those until you go to sleep every night. Keep in mind that the reel's fundamental use is a place to store line or apply drag to a fish. Otherwise, the line is in your hands or piled up at your feet or in a basket. Most fly fishers will agree there are several things you should get in the habit of doing when beginning to practice casting a fly rod. Here are some tips to help avoid some frustration:

- Assuming you have already tied a 6- to 9-foot piece of leader and/or tippet on your fly line, go ahead and tie on a fly you have clipped the hook from (something you do not mind beating up a little). The point of this is to duplicate the actual setup as much as possible for practice.
- With the rod in your casting hand and the reel on the underside, get in the habit of gripping the fly rod handle by placing your thumb on the top side of the cork handle just a skosh below the top edge of the cork. A "skosh" is defined as a little more than a hair, but less than a smidgen.
- If you can, get on the bank of a lake or pond, as it is best because of the "drag effect" water causes on the fly line. Otherwise, it is fine to go out in the yard or a field and

strip 15 to 20 feet of line off the reel and pile it up on the ground in front of you.
- If you have the fly line out beyond the tip of the rod, you can get this pile of fly line out past the rod tip by wiggling the rod tip back and forth, and the weight of the fly line will start drawing it out. This is where practicing on the water is even more helpful. The goal is to get all the line straight out in front of you (all past the rod tip), with no curls or loops, so when you cast, try to get all the line moving simultaneously. Make sure there is room behind you to backcast.
- Imagine looking at yourself from the left side and imagine the hands of a clock. As we proceed, try to keep the range of your casting arm strokes in the ballpark of 10 o'clock (in front of you) to 2 o'clock (in back of you) while performing your casting strokes (initial line lift off the ground or water excluded).
- Use your index finger of your rod hand to pinch the fly line (right after it comes off the reel) against the rod and hold it. At this point, your thumb and index finger on your casting hand are pretty much opposing each other, and the rod is pointing straight ahead at about 9 o'clock.
- Backcast the line from the 9 o'clock to the 1 or 2 o'clock position behind you and quickly STOP. Did you feel the load on the rod and see it bend to the weight of the line when you pulled it backward? Cast forward and quickly STOP at 10 o'clock. Feel the load again?
- Keep pinching the line with your index finger, and false cast back and forth, NOT letting any fly line slip out past the index finger. ***IMPORTANT: Feel the load (weight of the line) on the rod as it pulls the line in both directions. This "load" is your friend and will follow you all your days in fly fishing!***
- If you cannot "feel the load" so much, add some more line past the rod tip, and you will feel it. With additional line out past the tip of the rod, you will also notice you need to wait longer for the line to completely pass backward

and forward BEFORE you start going the other direction with your false casts. Try to avoid bullwhip snapping your line.
- As previously mentioned, add a target such as a loop of rope or just a spot to aim at and begin to get a feel for landing the fly near the target after a false cast or two. Let some line slip past your index finger as you forward cast to get the feel of shooting a little line out. Like any fishing cast, timing will come to aid in accuracy and distance.
- Practice *short* standard casts and roll casts, then go fishing to apply those skills on the water.
- As you gain confidence in your casting skills, consider using your non-casting hand to hold the line instead of pinching the line with the index finger of your casting hand. This will allow for "shooting out" more line during the cast and gain additional distance. Gaining additional distance via forward casting allows extra chances of hitting your desired target. Be patient as you work on this method. It will take a while, but it will also add to the thrill of the catch.

In summary, there are several tried-and-true teaching methods to help your casting technique. Fly fishing legend Lefty Kreh used the example of starting with practicing flipping the line back and forth on a flat plane in front of you like a windshield wiper, then practicing again at a 45-degree angle, and finally performing it vertically as if fishing. Acceleration is what loads the rod, but other instructors talk of "hammering a nail" or imagining "an apple on your fly rod" where the "sudden stop" of the forward or backward stroke is what launches the line.

Advanced Casting Methods

There are multiple other types of casting methods to assist in the delivery of your fly offering. These include single- and double-hauling to get more line out, curve and reach casts, and bow and arrow, tuck, and slack-line casts to get your fly in the right place.

A word of advice if you're a novice, it is still early in your fly fishing experience. Save learning these additional casts until you have more experience and a greater mastery of the fly rod in your hands. Do not think you can immediately go out and perform single- and double-haul casts. They are far more difficult, and you really do not need them yet. Once you understand the basic three casts, the additional skills for other types of casting will come easier. They will come in time, but not without considerable practice.

Distance casting in most saltwater (and some big freshwater lakes and rivers) will call for you to strip 50- to 75 feet of line off the reel and cast it all out and then strip it back in and re-pile it back in front of you, so it is in the correct sequence to be ready to cast back out when you see a school of redfish. In salt water fly fishing, spiked stripping mats help manage the line. You need to practice on smaller venues, tackle size, and double-hauling first. Spin fishing here is more forgiving, but with a little practice, fly fishing just catches more fish.

Remember that when saltwater fishing, wind and sun protection becomes even more important for the angler.

Finally, the basics of landing a decent cast on the water all end with a gentle settling. At this point, you are fishing, so be ready to react to any indication of a strike. Strip in any slack line as necessary to as little as possible to be ready to set the hook.

Fly Selection

Information on fly selection could fill a floor in your local public library. Entomology is defined as the study of insects. Multiple resources estimate there are over 10 quintillion bugs on our planet. That is eighteen zeros after the number 10. So, replicating each bug for those that get caught up on rocks and multiple hook sizes for each would lead to a tackle box the size of an ocean liner.

We are not going to go into the myriad of bugs a trout may eat. Many others much smarter than us have already done this, and the information is readily available for each specific location, circumstance, or scenario. Access the internet for various locations and water venues for a plethora of information. What will help is knowing the majority of any fish's diet is found under the surface, so "wet" flies like midges, nymphs, and streamers are good choices. Streamers are just lighter-weight imitations of the crankbait lures spin fishers use; you just "strip" the fly line back in instead of reeling them in on a spinning reel. Dry flies are exciting, but you will need to observe the activities on the water for a while to see what trout may be slurping off the surface.

For general selection guidelines:
- How deep is the water column you are fishing?
  - Getting down to just off the bottom is usually most productive.
  - Split shot can help get your flies down deep, and we have seen it work when placed above or strung out on tippet below the flies.
- Is the water still or moving?
  - Water moves faster at the surface than it moves down deeper. Using a strike indicator or even a dry fly (especially foam terrestrials) will drag the flies below it faster, keeping them higher off the bottom than you think. Use the rule of thumb of placing your strike indicator one-and-a-half to two times the depth to keep your flies down low.
  - Streamers are useful in bigger and deeper water as they pretty much are the equivalent of spin fishing with crankbaits. Judge your depth and add leader length if you are using floating lines to get down deeper.
  - One other suggestion: whatever streamer or wet fly you're using, change up the retrieve speed, location, or action before you change the fly.

- What kinds of bugs are living there?
  - Turn over rocks and grab a handful of weeds or algae and see what is living in the water. Imitate what you find with what you have in your fly box.
  - Dry flies are exciting to toss, but take your time to observe what the fish are actually eating. Do not be afraid to mix a buoyant floating fly on top with a midge or nymph dropper on the bottom. Often called "Hopper-Droppers," they are commonly used in spring and summer.

It looks like scuds and sowbugs are on the menu!

Our best advice for fly selection at any new venue is to talk to the local fly shops, game wardens, and conservation officers in the area and ask them the classic fishing question: "What are they hitting on?" We were fishing one day and getting no bites. An older man took pity on us, walked by, and simply said, "red midge." It made our trip.

Rick and I married twin sisters. Since they are Kentucky girls and raised accordingly, they are very accommodating of our outdoor adventures and fish with us or just go for a boat ride. Oddly enough, we have a habit of making sure we have a chance to fish on our anniversary trips together. On a trip to Ireland and Scotland, we caught brown trout at Ashford Castle (think *The Quiet Man* with John Wayne and Maureen O'Hara) and spent some quality time in Scotland. One day, Rick just had to play St. Andrews Links, so while Rick was golfing, Wade went to one of the local lakes to fish. Wade threw everything he could with no success. An old Scot showed up with his Labrador and was not there five minutes before he had caught two or three. He saw Wade's frustration, waved him over, and handed him a purple nymph. "Try this," he said in a thick accent. First cast and Wade had brownie in the net. Wade had to throw the ball for the black Lab a few times while the Scotsman fished, but it was a good trade. Takeaway? Fishing relations are international. Help your fellow fisher!

Build your fly box up from the basic wet and dry offerings mentioned in Chapter Two, curating it to fit the places you frequent and the fish you chase.

Setting the Hook with a Fly Rod

Holy cow, you have made a decent cast out into some fishy water. You are in the hunt now. You have a great, almost drag-free, drift going.

You have limited the slack line to a minimum.

Did you just get a nibble? A bump? A vicious strike?

There is one way to find out. Simply lift the rod to take the slack out. Still got the line pinched against the rod with your index finger, right?

Lifting the rod is all you really need to do. Well, and be ready for the strike.

Maybe you could accomplish the same thing by stripping in some line, too. In saltwater fishing, strip setting the hook is pretty much standard. In fact, if you are on a river or stream with a bunch of trees and brush behind you, a strip set of the hook might be your only option, and is why you need your free hand ready to assist at all times.

Once again, we will remind you the fly rod is a priceless piece of equipment. You shopped long and hard before handing over cash for the rod, right? OK, so let the rod do its job.

Here, we usually see the testosterone work against you (guys) and where the ladies usually have an innate advantage. Do not yank the line like all those bass fishing shows on TV! If you are a longtime bass fisher, these habits are hard to break.

Yanking the line or trying to muscle the fish will result in the fish simply having a sore lip as you watch him swim away.

Let us give you an example of a big trophy fish we once landed. We were tarpon fishing in Boca Grande, Florida. We had hired a guide, and the Captain and First Mate settled us in on some good water. The guide came up to us, lightly tapped his pinkie on the tip of the big baitcasting rod, and told us the little bump is what a 150-pound tarpon feels like when it hits. The guide told us, if you feel the bump, "Crank, don't yank."

He was trying to tell us to just take the slack out of the line. The hook will do its job. Tarpon scoop up food with their bucket mouth and put slack into the line you have to eliminate immediately. It worked. We caught a 185-pound tarpon.

Finally, if you are fishing on a small creek or have a small amount of line out and get a hit, just lift the rod with the intent of backcasting and forward casting to where you want to be next. One of two things will happen. You will either set the hook and have a fish on, or you will load the rod, backcast, and then forward cast back to try again.

Fighting a fish

Fish on!!!!

It is important to exclaim this when you hook a fish. Others near you will get their rods/lines out of the way. And you will have the added adrenalin of all eyes being on you.

The rod bends and loads up. As you learned, the fly line is pinched on the handle between your forefinger and the rod. The

fish is pulling some line past your finger, and the slack line is being taken up quickly. One of three things may happen:
a) It is a smaller fish, and the strength of the leader/tippet will handle the fish easily. The fish may never even take enough line to begin to pull additional line from the reel. Many people just strip line back in with their non-rod hand and bring the fish in without ever using the reel.
b) It is a stronger/bigger fish, and the slack line is quickly taken out, and the fish begins to take line against the drag of the reel. Make sure the drag on your reel is not so tight a hit will immediately bust the tippet. Less-tight reel drag will buy you time to add drag if you need to. Remember, if you are not minding the slack line, it can easily get tangled during a big fish run.
c) Wow. You have a big one on the line! If some fundamental concepts are not observed, disappointment can quickly follow.
   - Try not to overwhelm your tackle. Lighter leaders and tippet with a bigger fish can be a balancing act. Set your drag accordingly to provide resistance but not hinder the tackle. **Let the rod do the work.**
   - It sounds goofy, but try to keep the fish in (under) the water while fighting it. You can avoid some of the jumping and above-water gymnastics of a feisty fish by keeping your rod at angles closer to 9 o'clock or 3 o'clock. This method does not result in pulling the fish to the surface as much as a 12 o'clock rod position will cause. A fish jumping out of the water is exciting, but it will overload the tackle, land on the tippet, and/or spit the hook much more often.
   - If a fish is stripping out line against the drag, you have a dilemma. How much tighter can the drag be enhanced, if any? If you have room, let the fish run within reason, but not too long where the fish suffers total exhaustion. What obstructions is the fish going to get wrapped around if I do not turn him? If the fish is downstream on or after the strike, you are also potentially fighting against the

current! The answer is to feather the limits of the tackle as best you can. If possible, go with the fish downstream to reduce line strain. Keep your rod at low angles and try to steer the fish into slack water. Your instinct may be to "horse" the fish to the shore, but trying to overpower a fish to shore is a liability. You must play a big fish to land it. On the other hand, try to avoid fighting the fish to the point of its exhaustion impeding its ability to recover. The warmer the water, the quicker a fish will literally fight to the death. Many states restrict fishing in warm water with "Hoot Owl" rules, limiting fishing later in the day. Follow these rules and do your part to protect our fisheries.
- We all have stories of "the big fish which got away." Losing a fish teaches hard lessons on the expectations of the limits of leader and tippet sizes, fly selection, and, most importantly, personal patience.

Landing a Fish

You are making headway on bringing the fish in. The fish is almost in range of the net. Now comes the big decision time. Are you alone or with a buddy? If you have a buddy with you, here is where they earn their keep with the net.

It may sound like we exaggerate, but netting a fish is a bit of an art form. It seems so simple, right? Just dip the fish up. No problem. Wrong!

Some things to consider in netting a fish:
- As with the fight, do not try to quickly "horse" in and net a freshly hooked and feisty "green" fish. It insults the fish, and it will probably streak off on another run just to show you they are still in charge.
- Fight the fish long enough to tire it out, but not too long where it cannot recover. A long fight can kill a fish. Also, light leaders/tippet may not survive a longer fight, so it is your call on when to get the fish to the net.
- Do not stick the net in the water and try to drag the fish into it. Imagine you are the fish, and someone is dragging you into a lion's mouth. What would you do? We would

- all run like hell with every bit of strength we have left! So will the fish.
- Unless it is the sixth Thursday in a leap year February, always try to net the fish head first.
- The angler needs to get the head up out of the water and swing the fish toward the net so the netter can dip it quickly all in one motion. Coordination with your netting partner is absolutely key at this moment.
- If you're by yourself, operating a 9-foot fly rod in one hand and a short-handled net in the other hand is a challenging exercise. Basketball players have an advantage here. Avoid getting your leader inside the tip of the rod, as another run by the fish will probably hang up and break the leader/tippet. Reach out with the net and stretch back with the rod. Keep in mind these long fly rods are designed to flex incredibly all the way from the butt to the tip. If you are netting your own fish, do not be hesitant to bring it to the net.

Etiquette (last but definitely not least)

Fellow anglers are usually the finest of God's creations. As Jeff Foxworthy once noted, **"Look at where Jesus went to pick people. He didn't go to the colleges ... he got guys off the fishing docks."**

So, please treat your fellow fishers with the same respect you would give Matthew, Mark, Luke, or John.

Whether you are boating or wading, the first rule of good manners is making sure you do not crash on top of other anglers (see social distancing in Rick's Introduction). If there is a special spot you like to fish on the river, lake, or creek, wait your turn. We have a particular honey hole near the Wolf Creek Dam in Kentucky. Often, someone else is fishing "our" spot. We do not throw on top of them. We fish elsewhere and wait until they decide to move on before moving into "our" spot. While you are waiting for your spot to open up, be bold and explore a few new areas to fish while you are waiting.

Also, if you are upstream of other anglers, try not to stir up sediment (as well as fish). When wading, this is not always easy to do, so just take your time and make your way gently to your spot. If the shore is limited, we ask if the person minds us going a good distance above them. We have been denied only once. You could tell the fisher was a beginner by the amount of line he was leaving on the water. We simply went below him and caught fish while he was skunked.

If somebody is in "your spot," take the time to turn over a rock or grab a handful of weeds to see what bugs are currently emerging. It is nice to take a break when your back is stiff, sit on the bank and just enjoy the quiet.

Please stay off the phone. Your smartphone should stay in a waterproof bag unless you absolutely need it. We are all there to enjoy nature. No one gives a rat's behind about what is happening at your office, and for crying out loud, no speakerphone. PLEASE.

Good fishing etiquette should be afforded to the other fishers around you *and* those conservation officers protecting the world in which you fish. Game and fish officers are there to protect and defend the very places where we fish and enforce the laws of your particular area. For some perspective, Wade spoke with his brother, Roy, a retired 30-year veteran game warden of the Florida Fish and Wildlife Commission, about the most common issues a game warden encounters with fishers. Here is what he learned:

1. The most common occurrence is when the game warden is spotted by an angler way ahead of time while the warden is some distance away. The game warden has already observed the angler(s) fishing, and when the warden arrives at the boat or shore-based angler, dead fish are floating in the water near their boat or where they are standing. Usually, the angler possessed too many fish or ones that are out of the slot limit (the wrong size to be legally kept) and dumped them when he saw the game warden.
2. Another odd scenario occurs when anglers were observed fishing but still willing to go to court and swear to the

judge they were not fishing. They had rod, reel, bait, and all the supporting gear, but no current license and/or trout stamp, etc.
3. Finally, game wardens are often called to investigate incidents involving injuries or deaths on the water. As we mentioned earlier in the equipment recommendations, WEAR a life jacket if you are in a boat. The inflatable-type life jackets create NO restrictions while you are fishing and only operate if you fall out of the boat or are knocked overboard. If you are in your waders, use your safety strap so they do not fill up with water and bog you down if you take a spill. And do not forget your whistle. Accidents can happen, but sober anglers taking precautions can avoid becoming a statistic!

Here are some classic pursuits for any fly fisher:

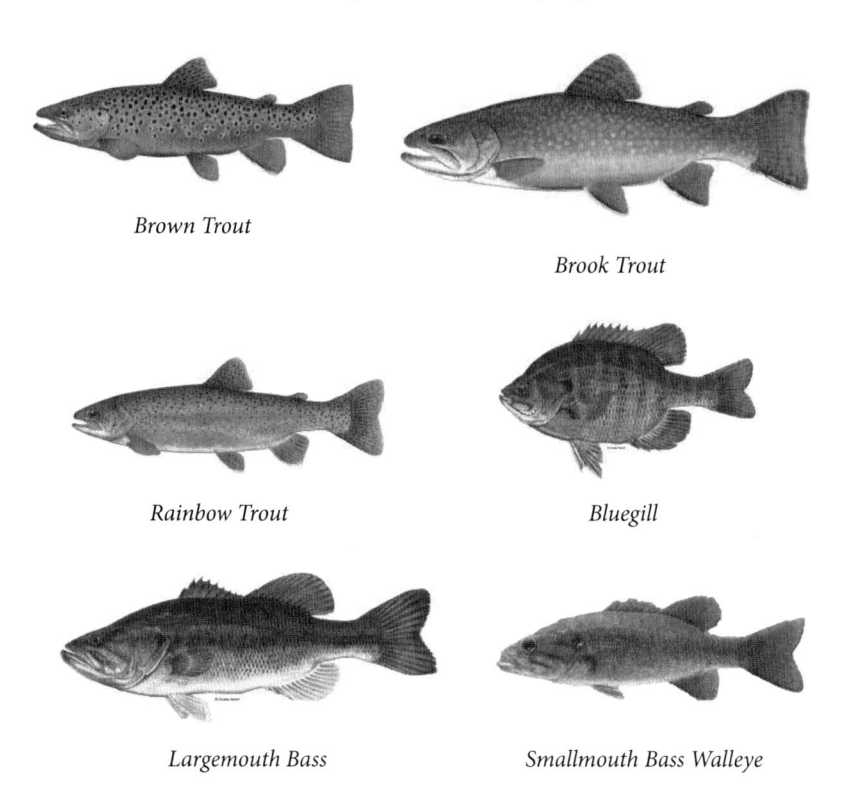

Brown Trout

Brook Trout

Rainbow Trout

Bluegill

Largemouth Bass

Smallmouth Bass Walleye

Photo credits: USFW

AND NOW ... A LITTLE HELP FROM OUR FRIENDS

How the heck do I use all this stuff? Now, that is one heck of a great question. A few choices as I see it: 1) probably the best thing to do is find a friend that at least knows the basics and have this friend start asking you questions and hopefully showing you a few things in the field; 2) watch YouTube videos -- you can seriously learn a lot about this sport from watching these miscellaneous videos; 3) local sporting goods stores (Cabela's, Scheels, Orvis) typically have some "how-to" clinics that are either cheap or free; and 4) hire a guide and either wade the river or float the river. I get more from the guide when we are doing wade fishing. Any time you are driving through a good fishing area, try to Google the local fly-fishing store. By and large, these are your best sources to find out what the fish are currently striking. This is not some -best-kept secret, but I have seen that the inexperienced angler is typically very intimidated walking into these local stores. It ends up that almost every single fly store I have gone to across the western states has been extremely helpful and eager to assist, so don't worry when you walk through those doors – be yourself and don't be afraid to tell them exactly where you are with experience.

—Leo Poggione
Reno, Nevada

The best tip I've learned and tried to pass on to new fly-fishing folks is learn to make a good cast. It really makes your time on the water more fun, less frustrating, etc., if you can cast pretty well. It's one less thing to have to worry about when you are hunting after fish. It's so much better when you aren't tangled up every other cast or hung up in brush or a tree all the time. I practice cast and encourage folks to practice off the water. That way, they can focus under ideal conditions to deliver a good cast and work out any issue. Having someone video your cast with your cell phone can make that even more productive. When you are on the water fishing is not the best time to start working on your cast!

—Pat Schleitweiler
Cincinnati, Ohio
Casting Instructor for Buckeye United Fly Fishers

Authors' note – As pointed out above, a person can fly fish for anything. Below are a few essays from fly fishers and their experiences at landing something other than trout.

"Now Suzanne takes your hand, and she leads you to the river. She is wearing rags and feathers from the Salvation Army counters. And the sun pours down like honey on our lady of the harbor. And she shows you where to look among the garbage and the flowers."

"Suzanne," by Leonard Cohen

Nymphing For The Native Cold Water Creek Chub Of Fish And Men: A Meditation On Unnatural Selection
By Mark Neikirk © 2021

I've been trying to decide whether the common creek chub is an ugly fish. It certainly is not attractive in the way the brook trout is attractive. It is not a sleek, hallowed missile of color and panache. Still, as fishes go, the chub rivals the bass, though in miniature, and is hands down more handsome than the catfish with its wormy whiskers, Apollyon skin, and trash receptacle of a mouth. Imagine the halitosis.

So mirror, mirror on the wall, which is the fairest fish of all? Maybe not the creek chub, but others are far uglier. If you find bulldogs cute, then the piranha is your fish with its ridiculously exaggerated underbite. Otherwise, it's no Prince Charming. The sturgeon gives us caviar, but there must be an inverse proportion in nature to exquisite tasting eggs and monstrously menacing looks. The Bass Pro Shop guide to fish (is there any higher authority?) describes the sturgeon as an "ugly lout with armored skin, a bulbous nose, a Hitler-like mustache of barbels." Not to bash barbels. The male chub sprouts little tubercles on its head during spawning season—a sort of two-day stubble, which looks good on George Clooney and, I guess, on chubs.

I am not claiming beauty for the common creek chub, *Semotilus atromaculatus*, or its kith, the river, silver, big eye, and hornyhead chubs. But I would like to claim some redemption on their behalf. Having gone to catch trout and catching only creek chubs, I have a stake in this. Those who would denigrate the chub might also denigrate me for catching them. So, chubby, we're in this together.

There's No Comparison. Or Is There?

A muse to painters and poets, trout are mythologized and monetized. They enjoy a status the lowly chub can only yearn to attain. There is no Chubs Unlimited to advocate on their behalf, no Orvis catalog dedicated to catching them, no vintage postage stamps of their visage for sale on Etsy. And yet, in the natural world, unstratified by human prejudice, the trout and the chub are equals in all the things that matter: What they eat. How they mate. Where they choose to live.

The diet of *S. atromaculatus*, whom I may henceforth just call "*S.*" was confirmed in a scholarly paper published in 1962 by the Iowa State Academy of Science, "Life History of the Creek Chub." It states: "Food items found most important for the creek chub were plant material, especially Cladophora sp. (algae); insects, with mayfly naiads and Coleoptera being the most important." Coleoptera are beetles. Mayfly naiads are immature mayflies also known as nymphs. Beetles and mayflies are dietary staples for trout. Chubs, too. Among the finned, trout and chub are similarly finical about their food.

What of mating? When chubs score, do they score on par with trout? Yes. Or I should say: Yes! Yes! Yes! The excited reply, however, needs context. The old insult of calling someone as cold as a fish when describing their inadequate heat in bed pales in comparison to the actual disappointment of fish sex. The female arrives at the nest, or redd. The male follows. Depending on the species, their bodies may touch. Eyes may roll, a fin may quiver. She releases her eggs. He squirts his milt over the eggs. It's a bit like visiting a broken ATM. No deposits. No withdrawals. The foreplay at least has a charming name: copulatory clasps.

Within the parameters of fish sex, a chub's Poseidon adventures are a notch above – assuming chivalry counts. For starters, the male chub brightens himself to a more kaleidoscopic spectrum, as if the date means something to him. The Peterson Field Guide to Freshwater Fishes describes the dazzle: "Breeding male has orange at the dorsal base, orange lower fins, blue on the side of the head, pink on the lower half of the head and body." The male chub also handles the nest-building, laboriously moving pebbles one mouthful at a time. The male brook trout leaves the nest-building – which takes days – to the female while he, quote, stands guard. The male chub does guard duty, too, but does so after the eggs are fertilized, when he uses his horned head to guard what his horny self has fertilized into zygotes. By this time in the breeding cycle, the male brook trout is an absentee father, on the prowl for another female. Scoundrel!

Chubs also let other minnows share the nest, which is not as neighborly as it might sound. The chubs appear to know that if predators come by, they might eat the other eggs first and spare the embryonic chubs. It's clever and also cinematic. Using underwater cameras, biologists have captured chub redds with schools of visitors swirling above. It's a feast of color and motion.

Food. Sex. Now habitat. Anglers love trout because of where trout live. Streams warmed and tainted by industry, mining, or sewage need not apply. S. is not so discerning. Its habitats were once fit for trout. By adapting when such places are spoiled by human indifference, *S.* is a vestigial reminder of what was. Were we still seeing creatures as gods, the god of conscience might be *Semotilus.*

While chubs can tolerate more pollution than trout, they do have their limits. Trout Park is a nature preserve 40 miles northeast of Chicago and so-named because 150 years ago it had trout. It also has creek chubs. Today it has neither. Interstate 90 construction and storm sewer routing fouled the water. In a cruel irony, biologists report that the preserve's "brooks still contain an unusual aquatic community including several rare species of caddisflies." Caddisflies are to trout, and chubs, what biscuits and fried chicken is to a southern preacher. How strange it would be to have a church picnic and no plump man of God there to consume the bounty.

My interest in chubs was awakened at Cane Creek on the outer edge of the Cumberland Mountains in southern Kentucky. Its story is a bit like Trout Creek's. God made it. We've ruined it. Cane Creek is as pretty a stream as any on the planet were it not for the local proclivity to trash it.

Particularly offensive are the plastic soda pop bottles. It is hard to swallow. Not the fizzy beverages necessarily, but the notion that we ask Middle Eastern kingdoms to drill for oil and, in an odd exchange. Next, we sell them modern weaponry to protect their extractive exploitation of their own land, then use the accumulated talent and enterprise of our century to convert the oil into easily held, aesthetically pleasing plastic bottles full of sticky liquids only the maddest of mad scientists could concoct. Potassium benzoate. Two kinds of citrate. Acesulfame potassium. Calcium disodium. Brominated vegetable oil. And dye. Double, double toil and trouble. Like a hell-broth, boil, and bubble. Then it is all handed over to the marketing folks, who mix their own witch's brew, a fantasia of celebrities doing some truly weird activities set to Wagnerian pop music pulsed by a drum machine.

The common creek chub has done nothing to degrade Cane Creek. It does not import petrochemicals from afar to make plastic bottles that will not biodegrade, even after an eon. It does not mix chemical compounds to fabricate a hydration syrup. And it certainly does not appropriate its celebrity to Cane Creek's detriment. In fact, S. has no celebrity – though I am trying to remedy that.

Not That Kind Of Nymph

There came to the Cincinnati Art Museum an exhibit celebrating the art of Paris, 1900. To see it was to be transported into the old City of Light's opulent world of cafés, opera, and galleries. The artists were A list: Renoir. Toulouse-Lautrec. Rodin. Claudel. Against this display of culture, our times seem deprived if not depraved. The Parisians of the Belle Époque strolled in splendor along the Champs-Élysées. We text and Tweet as we walk about wearing hoodies, smelling of Axe Body Spray, and counting our steps on digital wristbands as if putting one foot in front of the other were an achievement.

Among the paintings was one by Jean-Jacques Henner called "Recumbent Nymph." It is gorgeous; a little ghostly in its brush strokes but no less of a tribute to the female form for being a simulacrum in the old-fashioned meaning of that word, which was more celebratory of art imitating life. A boy looking at it would feel himself mature in its presence.

The nymph that trout (and chubs) eat is not that kind of nymph. It, however, also has an allure – though to fish, not boys. It, too, requires skill to execute. To paint "Recumbent Nymph" required exquisite ability with brushes, which at the time had bristles made of quill and stiff animal hair. To make a simulacrum of a stream nymph by tying a fly that mimics it demands skill, too, and is done with feathers stripped from a quill and the stiff hairs of an animal. The results generally are not found in an art museum, though the best examples might deserve to be there.

Tying a fly begins with a tiny hook. You might sooner find a dropped contact lens than a dropped Size 20 hook. All is minute except the task itself. After a foundation of thread is laid, feathers and hair are added according to a recipe, and that is the correct term for the roster of materials used to tie a fly, which seems appropriate since it is made to be eaten. Flies are small because they are meant to mimic the insects that fish like, particularly the small insects such as the mayfly, stonefly, and caddisfly that lay their eggs in streams. Those hatch into nymphs, which rise from the streambed to the surface to emerge as airborne adults. To fly fish means to match those hatches as closely as humanly possible by tying feathers and hair to look like the real bug, and to then cast the result upon the waters to test whether your artistry is adequate.

You can tie a dry fly to float on the surface or a wet fly to sink below. Since trout eat more underwater than above, nymphs make up more of their diet than adult insects, with perhaps 80 percent of what they eat being subsurface. In a stream virile with underwater bug babies, the fish forage. Hence, a fly fisher's arsenal includes an abundance of nymph imitations with names like Hare's Ear, Prince Nymph, Copper John, and Pheasant Tail. The names can get more creative: Flashback Scud. Zug Bug. Green Weenie. Picket Pen. Girdle Bug.

A Bigotry Most Foul

In his "History of Angling," Charles F. Waterman (love the name) takes note of what he calls the "strange polarization" among anglers. He then captures succinctly the status held by those who fish for trout: "The literary writing has been mainly about trout and trout fishing, much of it in moody flights of rather flowery prose." Averse to moody flights in his prose, Waterman is prosaic in recounting the brook trout's story, from abundance to scarcity. Humans – European humans to be precise since indigenous humans cannot be blamed – did irreparable harm, first with farms and factories but eventually with modern life in totality. Logging. Mining. Dams. Roads. Development. Plastic soda bottles. Reports Waterman: "The brook trout requires cold, clean water and was unable to compete with the early industrialization of the East."

The brook deserves better. The most elegant of trout, it is considered a grace wherever it is found. It is sleek and colorful. Our *S.*, smaller, more tolerant of gungy water, and born to the colors of *Casablanca* rather than *Gone with the Wind*, deserves better, too – especially in Kentucky, where there is scant evidence that the brook trout or any species of trout is native. One accomplished fisheries biologist admonished me to stop worrying about trout because the world doesn't need another word written about them. Turn your attention, he scolded, instead toward Kentucky's stunning aquatic biodiversity; a mile of healthy stream in Kentucky has more species of fish than in all of Europe.

He has a point. Trout are belauded in scholarship, in verse, in prose. "I am haunted by waters," Norman Maclean wrote in a passage that aches with truth as "A River Runs Through It" ends. Hemingway wrote about trout. Tennyson saluted them in "The Brook," where the water *chatters over stony ways in little sharps and trebles, then bubbles into eddying bays to babble on the pebbles.*

Over the centuries of human pursuit of fish, select species have been celebrated above others. In the sea, the honored include the swordfish, the sailfish, the Great Blue Marlin. Though neither is a fish, "Moby-Dick" and "Jaws" rank, too. Trout are atop the freshwater hierarchy, and sometimes the reverence is over the top. In Montana, I parked behind a Jeep, muddied by its life in the backcountry. It had a bumper sticker that read: "Trout Lives Matter." It was in the same typeface and design as the Black Lives Matter logo. Maybe the Jeep's owner thought it clever and meant no offense. Maybe that's worse, to be so isolated from the agony of being Black in America that you witlessly appropriate a slogan that is a clarion for justice.

Even by its insult of a name, the common creek chub has been subjected to bigotry, too. Bigotry toward fish is, by no stretch, a cultural sin equivalent to our bigotry toward each other. But it reflects something about our species. We are forever trying to elevate ourselves, whether by how we look or by what we do. I am white; therefore, I am. I fish for trout; therefore, I am.

Do we feel this superiority when we label some of the trout's brethren as trash fish or, in today's fish and wildlife bureaucracies, as Aquatic Nuisance Species, ANS for short? A species that abandons its bottles and cans hither and yon might not have the moral standing to attach "trash" to another species' name. Among fish, the reigning ANS threat is the silver, or Asian, carp. They jump out of the water en masse as if they were popcorn in a pot missing the lid. Now and then, one smacks a dodging angler in the face, fins slashing and gashing like an aquatic Freddy Krueger.

What happens above water is a mere spectacle compared to what happens below, where the carp's rapacious dining disrupts the ecosystem. They voraciously consume algae and bacteria that native fish need to survive. Bass get caught as babies in the silver carp's gills, a cruel end to a favored sports species usually destined to grow up and die after being reeled in by a human aboard a supercharged speedboat equipped with more electronics than a Tesla. Such a bass, if it measures up, goes on Snapchat first and then into the live well. If ever something was misnamed, it is the live well. It is a hospice where bass await death. Not just any death. To be filleted. Sliced from throat to belly. Neatly if the butcher is not an amateur. Most are.

If only bass could talk! They can sing. Big Mouth Billy Bass is so popular that a bar in Chicago installed 75 of them on its stairwell and rigged them to sing in harmony. When this Billy Bass choir collectively reflects on the live well, maybe it belts out "Whipping Post" or, as the night wears on and the anger wears off, fades to a Johnny Cash at death's door mood and moans "Hurt." By morning, the whole choir is feeling good about not being dead and breaks into a cheerful Bee Gees jive – "Ah, ha, ha, ha, stayin' alive, stayin' alive."

Our dear chub sinks never to such opprobrium. Nor do the other miniatures it lives among. Though not sized for anglers who gravitate to the waters' big and tall department, they are a fascinating lot. Shiners. Darters. Sculpins. Shad. Sticklebacks. A genus of small catfish, *Norturus*, or in common language, the madtom, is scattered and provides a roster of names that testify to the joy of those who got to name them: The tadpole madtom. The least madtom. The brown madtom. The northern madtom. And even the freckled madtom not to be confused with the brindled madtom.

A little-known sport has developed, micro fishing, built on a macro appreciation for the little guys. Micro anglers are the birders of the fish world, compulsively and impressively searching out species and subspecies to add to their lists and sharing each find on social media. The orangethroat darter especially caught my eye as I perused micro-fishing sites. It is green and orange and yellow, not just green and orange and yellow but the best possible shades of those colors. This is the company chubs keep.

Cane Creek, Where Chubs Enchant

I went to Cane Creek to catch trout. Arriving, I could see fish nipping at bugs. One's heart skips a beat at the thought of pending trout. Illusory, as it turned out. Chubs were feeding, not trout. Modern trout fishing is illusory in more fundamental ways. Cane Creek is a stocked stream and stocked under the charmless method known as put-and-take. Rainbow trout are fed nutrients in an array of aerated concrete tanks. Once they are six inches or so, they are loaded into trucks for transport. The truck comes up beside a stream that, if once fetid, has been cleaned up enough that these fish can live until they are either caught or die in summer when the stream gets too warm. A few find cool, deep holes and live longer.

Hatchery fish are more docile than their wild brothers and sisters. Even in the nineteenth century, as trout fishing took hold in America as sport rather than sustenance, anglers sought out wild trout. A hatchery is a fish factory. Underwater tubes shoot pellets of food to nourish fry into fingerlings. Where such fish are stocked, guides advise clients to use wet flies because the factory fish are conditioned to expect protein pellets delivered the way cheerleaders shoot T-shirts out of handheld air canons at ballgames. I learned to fish for trout not with a fly but with a kernel of corn that imitated a hatchery meal. I even made a corn lure, adding a yellow plastic bead to a silver spinner.

Think about all of this and ask again: Why are we hell-bent on loving the trout and not the chub, exalting one and disparaging the other? Never mind casting. Think caste.

I am not sure of the answer, having fallen in love with the trout myself, and especially the brook trout. The trout's affinity for clean, cold water is absolute, which means that you are going to an Eden when you go trout fishing. The fish itself is regal. Their torpedo bodies make them the Ferraris of freshwaters. They fight with a ferocity few fish can match. I once caught a brookie in a slender stream in the southwest Virginia mountains. Though no longer than a candy bar, the fish ran more patterns than an NFL tight end, slicing and darting in isosceles and then scalene. It flipped midair like Simone Biles.

In contrast, though superior to eat and much larger, the walleye barely resists being caught and avoids acrobatics. Properly fried, a walleye is succulent, with meat as white as Mitt Romney's teeth. But fresh from the water, it is as visually unexciting as next year's Camry. A brook trout's skin is a garden of wildflowers in bloom. A walleye's skin is the color of fallen leaves rotting in winter. A brook trout's eyes are angry. A walleye's eyes glassy. The countenance of one is fire, the other ice.

Precious few streams in Kentucky can sustain brook trout. With an irony you could not make up, the best of those streams is Bad Branch in Letcher County. Another is Parched Corn Creek in the Red River Gorge. There, in the 1960s, a lone fisher began his own stocking program, hauling brookies in his station wagon from Pennsylvania to Kentucky. He was breaking the law but also demonstrating a future possibility. Today, state wildlife biologists carry fingerling brook trout to portions of your state using state of the art transportation.

So much effort goes into the enterprise of making a stream into a trout stream that special regulations are developed to prevent anglers from diminishing the population too quickly. An extra license is required just to try and catch them. Creel limits and size limits are posted. In some streams, only barbless hooks are permitted. Certain months are off-limits entirely.

Creek chubs get no such love. *S.* survives on the strength of his, and her, own wit and grit.

A Debt Unpaid

Fishing is a pursuit. Someone catches, someone gets caught, if I might engage in some minor anthropomorphism with pronouns until I resolve who catches whom. We humans imagine ourselves the catchers. But in those most perfect of days, the roles are reversed. The fish catch us. We are smitten – consumed by the beauty of their world and trying to enter it with little more than a hook costumed in feathers and hair to look like a bug. We come to places as primordial as we can find within driving distance to escape the nation's deteriorated democracy and a world that values the virtual over the real.

When I arrived streamside on Cane Creek with a friend, he stood on a boulder and cast his fly into a pool below where I was walking. As I stepped into the stream, he called out. I looked over and could see splashing in the clear, rippling water.

A creek chub had taken his nymph. It was the first of several chubs we would catch on this day, a day the forecasters had written off as likely to be rainy. Instead, the sun shone, warming us to the point of shedding our jackets. No one else was fishing Cane Creek, so we had it to ourselves if we ignored the refuse of those who visited before us. Mostly, their garbage was out of sight, strewn above the bank beside the blackened logs and grey ash where they'd built campfires and left their mess.

Cane Creek is worthy of trout by everything the eye can see, though it cannot sustain them given the heavy metals, the acids, or other alterations that our use and abuse has inflicted on places like this. We are to blame, and yet the stream, and the creek chubs, have welcomed us here.

S., we owe you. Further affiant sayeth naught.

—Mark Neikirk
Crescent Springs, Kentucky

Endnotes

1. Any number of reference books offer insight into the variety of chubs. "The Fishes of Kentucky," published in 1975 by the Kentucky Department of Fish and Wildlife Resources, devotes 18 pages to them. *Semotilus*

atromaculatus is the star. While it is only member of its genus listed (most chubs are *Hyposis*), *Semotilus* can boast of broader distribution and abundance. "Statewide, in all headwater drainages," the state's book reports. The misfortunately named botched chub, or *Hyposis insignis*, seems to have gone AWOL from Kentucky stream in 1959. Oddly, *insignis* is Latin for conspicuous, which botched isn't.

2. The full title is "Life History of the Creek Chub, with Emphasis on Growth." The article was the work of James J. Dinsmore, later a professor and then a graduate student at Iowa State University. It is among the published Proceedings of the Iowa Academy of Science in 1962. JFK was in the White House, Bob Dylan was a Joan Baez understudy, and Coca-Cola came in a 6.5-ounce returnable glass bottle, cost a nickel, and was unavailable in Diet or Zero. A lot has changed since, but not the creek chub's diet.

3. The great biologist E.O. Wilson has reflected on the majesty of his discipline – a majesty that derives from how little is known about living things. So much remains to be discovered! Chemists can look at a Periodic Table and say, "Our work is done." We could travel the universe and not find any more combinations of atoms." Newton more or less nailed physics for us, and Einstein mopped up. But biologists are still discovering species and, more importantly, species interactions. More is unknown than known. What's true for biology overall is also true for fish sex. No one's precisely sure what happens with boy meets girl and fins flutter. The basics are understood, and for brook trout are well-described in the brook trout bible, "Brook Trout: A Thorough Look at North America's Great Native Trout – Its History, Biology, and Angling Possibilities" (1997, Skyhorse Publishing) by the late Nick Karas. I owe much to the Karas book, both in understanding this favored fish and framing the central question of my essay: Why the brook trout? Why

not the chub? Pity the chub has not such seminal work. To get my facts right on chub sex, I consulted the book mentioned in the essay (the full title is "Peterson Field Guide to Freshwater Fishes of North American North of Mexico, Second Edition," by Lawrence M. Page and Brooks M. Burr (Houghton Mifflin Harcourt, 2011). Credit also goes William M. Clay's "A Field Manual of Kentucky Fishes" (1962, Kentucky Department of Fish and Wildlife Resources) and, more personally, to Eric Hallerman, Ph.D., a fellow at the American Fisheries Society and professor at Virginia Polytechnic Institute and State University, who answered countless emails. Any errors belong to me as I tried to comprehend the incomparable intricacies of the miracle of life as the books and Dr. Hallerman tried to explain them to be.

4. For more on the challenge of understanding fish sex, and how chubs share their nests with neighbors, I found an article in the Transactions of the American Fisheries Society helpful. It's called "DNA Barcoding Elucidates Cyprinid Reproductive Interactions in a Southwest Virginia Stream." Dr. Hallerman, mentioned above, was a co-author along with his colleagues Brandon K. Peoples, Pearce Cooper, and Emmanuel A. Frimpong. They meticulously observed chub nests and analyzed the various species using them to tell the story of what the researchers called the chubs' "nest associates." Charles F. Waterman published "A History of Angling" in 1981. His book is, among other things, a stolid companion to the classic text, "The Compleat Angler," written by Izaak Walton, whom Waterman rightly crowns as the patron saint of fly fishing. Walton lived, wrote, and fished in the 16th and 17th centuries. Both books are a lesson on fish and fishing. Walton's prose is to Waterman's as a wedding dress is to slacks. On at least one occasion, the former is appropriate; on most days, the other is practical. It was Walton who elevated trout into the pantheon of fish. Everyone knows that. What's less known about "The

Compleat Angler" is that, in it, Walton may also have doomed the club to second-class status. As Walton's alter-ego, the teacher and guide Piscator, suggests that if a chub is caught and kept, it be given to the poor, for it is "full of small, forked bones... the flesh of him is not firm, but short and tasteless." The chub never recovered from Walton's slam.

5. Some biologists are certain the brook trout was never native to Kentucky; others believe it was, but, unfortunately, they cannot prove it. I've looked but have yet to find evidence of brook trout in the 1800s when their habitat would have been ravaged by logging.

The authoritative "Brook Trout" by Karas includes an appendix with state-by-state histories of the original distribution of brook trout. There is no entry for Kentucky, though there are entries for neighbors and near neighbors, Tennessee, West Virginia, Virginia, North Carolina, and Pennsylvania – all places where strains of native brook trout remain. Even Ohio, with no brook trout fisheries today, has an entry based on the likelihood that it had brook trout at one time. Illinois and Indiana have entries, too, not because they have brook trout today but because they probably once did. Of the seven states that border Kentucky, only Missouri is ignored along with Kentucky in the Karas appendix. Still, there are hints of a native brook trout history in Kentucky from other sources. Fish surveys in the early twentieth century suggest population in the higher elevations of the mountains along the Virginia border, but the evidence is inconclusive. The brook trout's authenticated history in Kentucky begins in the 1960s, when a sign painter from Louisville, Bill Holmes, began smuggling them from Pennsylvania to Parched Corn Creek in the Red River Gorge, creating his own brook trout fishery. Writer Kevin Kelly told the Holmes story in the Winter 2015 Issue of *Kentucky Afield* (pp. 12-15). The Holmes story is a love story between man and fish, but it is forbidden love. Stocking a non-native species is against the law. But, as Kelly reports, by 1980, it was the state Department and Fish and Wildlife that was bringing in brookies. Bad Branch in Letcher County was among the first steams stocked. Bad Branch is off-limits to fishing; Parched Corn is open to fishing though any brook trout caught must be released.

Once, since I have worked as a tour guide here in Iceland for eight years, I had many Americans from all over the U.S. on board. One guy once asked me if I believed in God. I said, yes, sir, I do, and my touch to the Almighty is through the fly rod and my fly line when the fish strikes. That moment is heavenly for me.

I mostly fish in still water, so these tips are about still water fishing.

1. Fish short casts first simply to spook nearby fish, not the fish that is farther away. Long casts are a great ability to have, but most of your bites will take place once you have already dragged in at least one-third of your line and sometimes more than that.
2. Fish the banks! There are so often fish ten feet away from the banks, and skipping fishing the banks is like skipping a bite, I'd say.
3. Fish 180° angle, start from one side to the other and scan the area if you don't see any fish.
4. Don't dwell for too long on one spot if nothing is happening. Be mobile and move; scan the water with flies like Mickey Finn and other good patterns that work great to search for fish.
5. Most of the anglers often pull up their line too soon. Sometimes a fish is chasing your fly without you knowing it, and it may just take it when the distance is only about 1-2 rod lengths to it.
6. Mix up your drag! If one doesn't work, mix the strip up. Try fast and slow, try moderate speed, and all of those three mixed just in the same cast.

Last advice, fishing isn't about catching fish only; it's about being out there, in nature with the flowers, bees, trees, birds, and all that. It's for most people about charging the batteries.

—Ívar Örn Hauksson
Iceland

Know your Knots:
Keep it simple. Know three knots.
1) Fly to leader/tippet = improved clinch
2) Leader to tipper = double surgeons
3) Nail knot (purchase the tool that doubles as the nipper) is best recommended for line-to-leader, so it does not get hung up in the top guide. Plus, it's very strong!

—Mike Masciangelo
M&M Trout Adventures serving
Missoula, Enis, and Bozeman
Check them out at www.mmtroutadventures.com

The environment is a puzzle composed of many pieces like current, tide, wind, topography, predator/prey relationships, and so forth. You can learn some of those things from books and some from friends or guides, but mostly you have to put some time in on the water. When I first started fly fishing in saltwater, I probably dragged myself out to the bay a dozen times without really catching much. But that time on the water paid off because I came to understand at the right place, at the right time, something usually happens. There's a sort of point of intersection. When it all comes together, you'll know it.

—Jack Keller
Bayview, Texas

What a Grand Thing It Is; a reflection of 24 June 2020
By Mark T. Sorrels © 2020

Following a ten-hour day of work, in which I sold two new commercial-grade Maytag washing machines, I made my way home to South State Street. At the beginning of the day, I rejoiced in my soul from the fellowship I shared with the Holy Spirit and the encouragement He gave me to present my body and mind as sacrifices to God so that I may discern His perfect will. That time of connecting with God set the tone for my attitude throughout the day. What a grand thing it is to know and serve the living God who sent His Son to buy me back from the destructive work sin had done in my life.

Arriving at home at the same time as my wife, I made my way to the kitchen and found the rice and lentils she had prepared to be tasty and filling. She prepared them for her lunch and earlier had phoned to tell me there was enough left for my supper. What a grand thing it is to eat a delicious meal prepared with love by my spouse.

After supper, I gave thought to go fishing. But, as my sense of being responsible overtook the desire to fish, I fetched the string trimmer and tidied up the edges and fence row of the lawn. What a grand thing it is to have a house with no mortgage and the physical ability to care for it and the lawn.

The clock said 7:20 p.m. Knowing sunset was not until 9 p.m. or later, I got in the truck and drove to a familiar spot of the Muscatatuck River. With fly rod in hand, I negotiated the overgrown path that led to the river. The water was slightly stained and running wider and deeper than five days earlier. I studied the currents and planned how and where to fish them. Having twice fallen, I was aware of the "danger zone" of rocks and boulders in the riverbed. Minnows broke the water, and small birds flew like fighter jets inches above the surface. Using a small crawfish imitation, I landed a small but beautiful "brown fish" from behind a large boulder in the middle of the stream within ten minutes. After another 45 minutes of slowly moving upstream and catching nothing, I spent another fifteen minutes fighting a high limb of a mature Sycamore tree to save my bait hung in its leaves from an errant backcast.

After rescuing the crawfish imitation, I tied on a self-made, olive-colored woolly bugger and continued moving upstream. Ahead of me about 100 yards was the tail of a pool from which I seldom caught anything. Still, it always looked like good smallmouth habitat to me. As I waded into the slack side of the pool, the sound of the rushing water in the riffle below drowned out the arrival of two young adults. I became aware of their presence as I saw them crossing at the riffle out of the corner of my eye. On the other side of the river, motivated by hormones, the couple acted as if I was not present. As I continued fishing, twice I heard the guy's cell phone ring.

At approximately 9:10 p.m. as my woolly bugger drifted with the current just inches from the bottom of the riverbed, I felt a powerful thump in the rod and saw the line move upstream. I set the hook, and the drama began. Hooking a big smallmouth on a No. 5 fly rod and line is a thrilling experience. My right hand pulled in line as the first finger of my left hand served as a lock to hold it secure. However, the "smallie" had other plans and more power than I expected. He then moved across the river and peeled off a large length of line in doing so. Slowly I began to recover line and move the fish toward me. Wanting to land the fish, I did not pressure it and take the chance of breaking my leader or pulling the hook from its mouth.

The fish and I fought like two evenly matched boxers in the ring, neither getting an advantage over the other. I would win for a few seconds; then the fish would have its way. As the battle continued, I yelled to the guy on the other side, "Hey, get out your camera." After a few seconds of watching, he said, "I think that's a pretty good fish." To which I replied, "I think it is." At that moment, he began taking photos of the experience. As I fought the fish, it swam downstream toward the riffle, using the increasing current to its advantage. I knew I needed to get the fish away from the riffle. Using both hands on the rod, I managed to get it back near where the drama began. That big "brownie" jumped entirely out of the water two times, trying to shake the hook from its mouth. At last, the brute began to tire after a long and gallant effort to free itself. I brought the fish to my thumb and forefinger and lifted it out of the water long enough to remove the hook and to allow the guy with the camera to see it.

I looked with amazement at the wonderful creature as I gently placed him in the water and supported him for a what may have been 40 seconds or more. As I watched, I realized I had seen its color change during my experience with it. As I fought and landed the fish, it was golden-brown. When I placed it in the water, it turned olive with black bars on its sides. As I watched this beautiful fish recover, I stood amazed at the beauty of nature and the creative hand of God. Regaining strength and equilibrium, the "Muscatatuck Monster" slowly swam out of my hands back to the depths from which it came. I gave the photographer my phone number and he sent me 29 photos covering a large portion of the action via text. I looked at the guy and said, "I think it's time to go home." What a grand thing it is to have a river with smallmouth bass less than ten minutes from my house.

What a grand thing it is to know Christ, to have a Godly, loving wife, to have a mortgage-free house, to fish in a beautiful setting less than ten minutes from my house. What a grand thing it is! What grand things these are!

Our Family's Greatest Catch

Many people tell stories about the biggest fish they ever caught or the one that got away. This is the story of our family's greatest catch.

My son, Logan Bliss, is an avid fisher. Mostly bass, but also muskie, walleye, and trout. He even went to the University of Wisconsin, Stevens Point, known for its natural resource school and fishery management classes. As a founding member of the UW Stevens Point Fishing Team, he brought in prizes from Texas, California, and many other states. His first career was with the Wisconsin Department of Natural Resources Fish Hatchery, working at the hatchery and doing stream restorations.

One day he decided to go out in the early trout season, a catch-and-release season. He did not bring home trout, but he did bag a catch. This day, at a remote trout stream, a dog started following him for hours and would not go away. Being the kind soul he is, Logan looked at the dog with sympathy. It was visibly underweight, starved for affection, and very muddy. The dog had a collar and tags, so Logan called the number on the tag to let the owner know the dog was found.

The crabby old man on the line responded with, "Ya, I was going to get rid of her anyway. Do you want her?" Being a newlywed, Logan said he would have to ask his wife. His wife was so upset that this dog had been neglected and mistreated that she said to bring it home and they would give it a good home.

They nursed the malnourished, worm-filled, Lyme-diseased dog to health. "Teal" now lives in the lap of luxury with two loving parents and two skeptical cats that are unsure if they like the new addition to the family.

Only good things come from fishing. Sometimes you go fishing and come home with man's best fishing buddy.

—Amy Bliss
Cottage Grove, Wisconsin

The Great Alligator Hunt
By Rick Robinson © 2021

Prisoner's Lake is now well stocked with rainbow trout thanks to the Kentucky Department of Fish and Wildlife. However, there was a time when the lake also contained alligators.

My great-uncle Chip Thomson would spend his winters fishing and hunting in Florida's Everglades. When he came back to Bromley, his Jeep would be full of more than just bags of Indian River fruit. Usually, the extra load consisted of stories he would weave for the folks at the local bars about tracking wild game. On occasion, he would also bring back a snakeskin or a stuffed alligator head. One year, he brought back something a little more tangible—two alligators for Bill and Tom Gaither, the sons of his best fishing buddy, Rudy.

The Gaither boys were enthralled by their new pets and had every intention of keeping them. They dug a pit behind their house on Carneal Street and put chicken wire around the hole. They started charging kids in the neighborhood a nickel apiece to come and look at the alligators. A steady flow of kids—and adults—made their way to the Gaithers' backyard. The enterprise ended when Mrs. Gaither tried to push the alligators back into the pit with a broom and the larger of the pair bit the handle in two. The collected admission having not covered the cost of a new Fuller Brush broom, the Carneal Street Alligator Park abruptly closed its doors to Ludlow's animal-loving public.

Uncle Chip and Rudy decided to give the alligators a new home at the Behringer-Crawford Museum. They put the two reptiles in temporary cages and went about building more secure permanent quarters. On the third day of construction, Chip and Rudy showed up at the museum to find the alligators missing. A search of the museum grounds turned up nothing. The two men assumed the gators had been stolen.

It wasn't long until Covington police began getting phone calls about alligators roaming around the golf course. One evening, the reptiles wandered through a performance at the Devou Park concert bowl. After a quick investigation, the police knew whom to call.

Word spread quickly throughout the community about the escapade, and people began searching the park for the scaly escapees. They found them in Prisoner's Lake. About a hundred people lined the shore of the lake when Chip and Rudy showed up to catch the alligators.

According to an article in the *Kentucky Times-Star* titled "Big Game Hunt on for 'Gators in Devou," it was quite a scene. Chip went out on the lake in a canoe and lured the first gator to the surface with some raw chicken. When the first alligator emerged for the fresh meal, Chip was able to catch a treble hook under one of its stubby legs. Rudy stood on the shore with a square fiberglass Wright Magill rod-and-reel struggling to get the animal to the shore. Chip jumped from the canoe into the lake and onto the alligator's back when he was within range. He grabbed ahold of the animal's jaw. Rudy took some electrical tape and wrapped it around the alligator's snout. It thrashed back and forth as Chip and Rudy dragged it onto the shore and carried it to the pickup truck.

Rudy tossed the exhausted animal into the bed of an old pickup truck and then they went about catching No. 2. It was smaller but much quicker. It took Chip a bit longer to hook him. But once he was hooked, bound, and tossed into the back of the truck with his fellow traveler, a roar went up from the crowd. The men and boys tossed their hats in the air. Newspaper photographers caught it all for the morning edition.

The next day, the local newspaper published a photo of Rudy struggling with the bent fiberglass rod to get the animals to the shore. Rudy sent the photograph to the rod manufacturer, and they sent him a whole bunch of free gear. They used the photo in their advertisements for years with the caption, "Tough enough to catch alligators in Kentucky."

All the kids gathered around the pickup truck and got to touch the alligator's rough, spiny skin. Rudy and Chip told everybody that they were going to give the captured pair to the

Cincinnati Zoo. As an adult, I went to the museum and saw a stuffed alligator on display that looked awfully familiar. I always had trouble looking it in the eye.

So, when you're fishing for trout this year in Prisoner's Lake, think twice before remembering to catch and release. The alligators like trout, too.

A fictionalized account of this true story is contained in the novel *Alligator Alley* by Rick Robinson, Headline Books.

1. Always wear a wide brim hat and glasses. The guy that taught me how to fly fish told me this. He said that it was good protection from a fly that is coming back at your face. My very first time fly fishing alone the fly went right into my unprotected eye and stuck into my eyelid. I made it back to the car and used the mirror to remove it. Fortunately, no serious injury, except to my pride.
2. It is OK to tip your fly with a waxworm, but don't tell anyone. Remember to make sure it doesn't get in your selfie. I always have waxworms or PowerBait waxies handy to help with "matching the hatch."
3. Remember that the length of your passenger compartment or trunk space in your car is always one inch shorter than the length of your fly rod. That is the real reason that a lot of fly rods come with two tips.
4. The safe distance behind you so that your backcast will not snag anything is indefinite. You might not be able to cast 100 feet in front of you, but for some reason, it is easy to reach that tree 100 feet behind you. Like one time in my case when I snagged a stop sign in front of a bar by the bridge I was fishing off. . Fortunately, after I unhooked from the stop sign with three drunk smokers watching me and secretly tipped my fly with a waxworm, I caught a nice bass and redeemed myself.

—Bob French
Chicago, Illinois

4

So How Do I Go About Catching Some Fish with My Fly Rod?

"Listen to the sound of the river and you will get a trout."
—Irish proverb

The authors of this book are split on where to fish. Because Wade lives on the Cumberland River near a spot called "Rainbow Run," he prefers river and stream fishing. When you can walk out your back door to fish, it is too tempting to choose otherwise. "I love a river setting more than any other venue," Wade declares. "Lakes and oceans are fine, but it is an easy choice for me. I believe a river is like a unique signature of the country I am in at the time. The river says you can visit today, but I will be a little different next time."

Because Rick lives near several state-stocked lakes, he spends most of his fishing hours on still water. "I like fishing for a variety of fish," he counters. "The lakes by me are stocked with rainbow, but I get just as much of a kick out of landing a crappie or a smallmouth on a fly rod. Although, the toughest fight I ever had was when I landed a catfish on the Potomac River from the Virginia shore."

This contradiction between the authors points out one myth about fly fishing. It is not just for trout – you can fish for anything on a fly rod.

Wildlife artist Bill "W.D." Gaither loved to fly fish off Galveston for tarpon, and his favorite fish to land was a smallmouth bass. The authors having once linked into a 185-pound tarpon with a big bait and cast rod cannot begin to imagine how Bill caught a tarpon on a fly reel, but he did so. Of course, Bill would never tell a wild fish story … never.

In Florida, there is a fly-fishing tournament for sharks. While we would love to watch it sometime, the thought of wading waist-deep through chummed-up, shark-infested waters is a little too much excitement for us. Last year's winner was named "Stubby."

The Potomac River story mentioned above is so absurd, it bears telling. A local Orvis store salesperson had told Rick one of the best places to fish for smallmouth bass was a broad stream less than 100 yards north of the runway at Ronald Reagan International Airport in Alexandria, Virginia. The stream was short and ran into the Potomac River. Rick bought a couple of red worm flies, grabbed an Uber (yes, an Uber, it was D.C. after all), and headed to the stream. There was a boat ramp and a dock near the intersection of the river and the stream. Rick wandered out on the boat dock and started tossing a line. Nothing.

"Then," says Rick, "I tossed between the ramp and the shore. Something hit it and took off like a rocket. I immediately thought I might have stumbled onto one of the legendary stripers on the Potomac, but those usually hang out a few miles further upriver."

Job one was to direct the yet unseen fish to run towards the Potomac River and not allow it to escape under the dock. Once the fish was headed out to the river, Rick turned the fish and himself back to the stream. But when he turned, he realized he was not alone. A yellow duck boat filled with foreign tourists had stopped to witness the legendary fight. So as Rick struggled with the fish, the boatload of tourists took pictures. The guide called play-by-play in some unknown language, his voice rising and falling with each turn of the fish.

"When I finally got the fish to the top," Rick said, "I realized it was a pretty good-sized channel cat. Of course, I netted it and held it up to the cheers of the duck boat patrons."

In the end, time and place matter not. In the words of Canadian conservationist Roderick Haig-Brown, **"There will be days when the fishing is better than one's most optimistic forecast, others when it is far worse. Either is a gain over just staying home."**

Reading the water

No matter where you end up wetting your line, it is important first to read the water. Journalist Tom Brokaw once declared, **"If fishing is like religion, then fly fishing is high church."** The river or stream becomes the holy water of the religion of fly fishing."

Leonardo Da Vinci understood the connection between water and life. **"The water you touch in a river is the last of that which has passed, and the first of that which is coming; thus, it is with time,"** he said.

You will hear all fishers talk of reading the water. And while it has a kind of cosmic sound to it, the phrase simply means: "I wonder where the fish are?" The answer often depends on the type of fish.

Let us start with the elusive trout. Trout are basically like people; if they can find an easy way to get something accomplished, they do so. So is the way they seek their food. The less energy expended getting groceries, the better. You do not stand up when you're watching TV, do you? I'm going to be in a lounge chair chillin', hoping for a sandwich to come drifting by. Humans do not have a monopoly on this concept, and you should expect fish to do the same, so fish lazily for trout.

Many times, you may hear the mantra "foam is home" for trout anglers. Trout have a habit of hanging out in the "seams" of creeks, lakes, and rivers. Seams are where moving water meets still or calmer water, and foam usually marks this confluence or "seam" in the water. Trout hang out there because they can take it easy at the edge of the slower current and wait for food to drift by in the faster current. It is kind of like a reverse drive-thru or buffet where the food moves instead of you.

For the same reason, trout are also fond of cover. Trout will often hang out just in front of rocks or a few feet behind them because the current causes less body drag during their eating routine. Watch for the "soft water" in front, sides, and several feet behind rocks or other structures, and you will come to understand it is a great place to find fish. "Soft water" refers to pockets of slower and/or deeper water surrounded by faster currents.

In other venues, such as where the water may be deeper or shallower, faster or slower, or warmer or colder, fishing tactics may change because the pickings for the fish are better somewhere else. Depth changes are common and can be subtle. Scan the water for the changes in the bottom level (especially rivers) and structures for lurking fish.

Perhaps a hatch of flies is occurring. Bugs (larvae/midges) emerging out of the mud on the bottom of the creek lake, or river are easy marks for a hungry trout. Nymphs rising through the water column headed for the freedom of the surface are easy targets. Land-based food sources are often abundant at certain times of the year. Grasshoppers, beetles, ants, etc. are referred to as "terrestrials" and one landing at the right spot may be just what the fish is seeking. Finally, the surface of the water may be littered with newly emerged flies attempting to take off into flight only to be slurped into oblivion by surface-feeding trout.

d the inhabitants of the watery element were
men to contemplate, and fools to pass by
ration." - Izaak Walton

...ngs us to the next choice you will need to consider…

Presentation

What is your fly selection today? Wet flies? Dry fly? Streamers? Presentation affects success with all selections. As we have pointed out, flies are just imitations of food for a fish, the same as spin fishers use crankbaits that float or sink. You are just using a different tool to deliver the illusion of food.

What a Drag it is

Avoiding drag is a huge key to success in fly fishing. We have written about casting and fly selection, but without emphasis on keeping your fly drag-free, you may get skunked. A wet or dry fly being dragged by the fly line just looks unnatural, and any self-respecting fish will reject your offering. This is why combining the correct cast and fly placement is important, as well as quickly mending the line after a cast to extend the "drag-free drift" as long as possible.

Drag happens both on top and underneath the surface, but the water flow is much faster on the surface than deeper down. This affects the realistic appearance of the fly to the trout. Moving water has a flow effect on your fly. If the fly's travel is affected by line drag, then there is a fat chance of your fly getting sampled by a tight-lipped fish. *Strike indicators can add to the drag problem,* as can the type of knot used on the fly. Think of it this way. If you were to just throw your fly into the water with no line attached to it, how would it float and drift? Try to mimic the same motion when the fly is tied to your line. Knots like the mono loop or similar let the fly swing free and can lessen the unnatural movement of a fly.

Here are a few ideas to use on the water. If your chosen venue presents a "foam is home" seam in the water, your presentation can be accomplished in a few different ways. When wading, casting to an area upstream of the desired or sought-after spot where fish may be hanging out can give your offering time to settle and travel through the desired area. Depth of the water, use of a strike indicator (or not, try both), and of course, changing flies may be part of the search to get a fish on the line. If you're nymph and/or midge fishing, try to get your wet flies down to within 12 inches of the bottom. A good drift will be drag-free about 6 to 8 inches off the bottom. This is right in the wheelhouse of trout and other fish and will take some practice and adjustment to accomplish. If you are not bumping the bottom, you are not down far enough. If you bump the bottom to the point you are getting hung up a lot, adjust the strike indicator, lift the rod a little or do a little strip to get the fly off the bottom. Manage your fly. You will begin to feel the difference between bumping the bottom and a strike. Again, work on the presentation issues before you decide to change the fly selection.

If you are in a boat, it probably is even easier as the drift of the boat takes care of much of the need for "mending" of the line. Fishing guides commonly use drift boats and river fishing in this method may allow the coverage of multiple "river miles." This type of fishing venue will give you a greater perspective of the different types of presentation opportunities. Again, listen to your guide.

As an example of not listening to your guide, Wade was fishing the upper Columbia River in Washington with Evening Hatch guide Rial Blaine in 2017. He had not caught a fish in a while during the drift, most likely because his presentation was not consistent with local knowledge of what works in the area. After Blaine watched Wade for some time not doing what he had suggested at the beginning of the trip, as tactfully as he could, he asked, "Now Wade, remember what we talked about?"

If you are dry fly fishing, you have chosen a fly imitating the real flies you observe on the water and around you. The challenge is to consistently reset the drift of a dry fly to present a realistic food option for a trout. A couple of "dead drift" presentations, where the fly flows with the current, can be created by casting either directly upstream or downstream. In an upstream cast, the fly drifts back to you and requires you to strip slack out of the line as it does so. With the downstream drift, the initial cast is shorter and you feed line out as the fly floats down with the current.

A couple of tips come to mind. Fly line tends to bond to the surface of the water and makes a slurping snap when lifted off the water to backcast and forward cast to the new drift. This is noisy and disturbing for any fish but can be reduced and/or defeated in a couple of ways.

The first and easiest thing to do is lift the line and take up as much leader as possible with a mending action. This is a preparation to the backcast and sets it up without the ruckus of just jerking the line off the surface in one motion. Another preparatory advantage to avoid the bonding of the fly line to the water (causing slurp/snap) is to use an old firefighter's tip. Firefighters use foam as an extinguishing agent called a surfactant (sort of a fancy name for detergent). Two things happen: When you wash your fly line often, it removed contaminants bonding to the fly line and affects casting and slipperiness (dirty line through the rod guides), and by washing it frequently with detergent (surfactant), a layer of soap film helps the fly line to not bond to the water. End result? Smoother casts, less water surface disruption, and longer drag-free drifts.

Jack Mitchell at *The Evening Hatch* recommends "**less line is better**" and you should make a preliminary cast 50% of the way to the target to get an idea of the challenge. Shorter casts are easier to present properly, especially with slack created in the cast to increase drag-reduced drifts. "If you think you are getting a drag-free drift, think again. The only perfect drag-free drift is a fly drifting down a current lane with no leader/line attached. If there is a line and leader attached, there is drag; the question is, to what degree?"

Keep in mind you will not always see the strike on a dry fly. It is beautiful to observe, but the type of fly and where it is riding might not allow it. The fly may be just under the surface, so when in doubt of a strike, set the hook. If nothing is there when you try to set the hook, remember to complete the movement as if you had intended to recast.

If you are fishing streamers, look for the route you want to drift or strip the imitation bait back in. Get in a position to make a cast to achieve the retrieve you are seeking. If you keep casting often from positions where you do not have the best advantage, you are just disturbing the water and putting fish on alert rather than offering a food opportunity. Also, change up your stripping in (presentation method) before changing streamers.

To sum up, on presentation options, do not get hung up on the idea that big bait = big fish. As big fish can handle big food options, sometimes this is true. Yet, you will inevitably see a small fry attack lures or flies as big as them. Conversely, we have also caught big trout on small flies many times.

To add some options, Wade is a fan of mixing it up with dropper flies off the main course offering. Try a woolly bugger (10 or 12) with a straggler small prince nymph behind it and vary your retrieve. Next cast let the heavy woolly bugger settle and the nymph flutter, twitch, and strip and try settling again. The worst case is you get some interest and try just a nymph tactic next.

We are firm believers you can be in a spot where you can catch a lot of fish, or you can be in a spot where you can catch big fish, but not usually both. Now do not misinterpret the statement. We appreciate catching any fish. But as you evolve in your fishing expertise, you may find yourself targeting places where larger fish hang out. If someone asks you at the end of the day fishing, what would you like to say? I caught ten little ones or one trophy? Your call.

Shifting Gears

This is as good a place as any to point out that the authors are not fly fishing purists. Dare we say it? We actually also go fishing with spinning rods and crankbait. Yes, we go both ways. One of us occasionally, oh the humanity, uses worms to catch fish. Gasp.

Quite honestly, there are times when fly fishing is practically impossible, and we try never to miss any chance to wet a line. Period.

Several years ago, the authors took three young college boys for a drift down the Cumberland River. The water was perfect – so skinny we had to get out now and then to push the boat forward. While it would have been ideal for fly fishing, running five fly lines around a 17-foot boat would have left us spending more time untying tangled lines than fishing.

With an accompaniment of every color rooster tail imaginable, we put in just below the dam. Ten miles later, the five of us had caught 137 trout and one striper. Each of us hit the trifecta – a rainbow, a brookie, and a brownie. All were quickly caught, gently wet released with minimal if any handling, and lived to fight another day.

As we were adding up the numbers between the five of us, one of the young men exclaimed, "Wow, just imagine if we had used a guide." This has now become our standard retort when we have a great day fishing.

We point out our sin of spin casting, not to ask for forgiveness. Come on. We have all done it. We simply ask, when fishing for trout, fishers add a couple of rules to your routine.

1. Avoid **treble hooks** – cut one (or even two) of those off and the bait swims just the same and will cause less harm to a hooked fish. There is an argument both ways about this, but it has been our experience that treble hooks can cause far more damage. See our "Catch and Release" segment.
2. **No barbs** – almost all the hooks on spin bait have barbs. Clip them off or crush them down. This goes for flies too. Remember, tight lines will negate the need for barbs. Many folks believe a barbless hook will penetrate deeper and be less likely to lose a fish if the line is tight!
3. **Release** – Try to release the trout by quickly flipping a pair of needle-nose pliers at the curve of the hook. It does not take long to learn how to release a trout without even touching it. The less handling, the better, and keeping them in the net and in the water the whole time you are dealing with them is even better. At the local hardware store, you can find needle-nose pliers with a long handle and the nose bent at a 45-degree angle. These are perfect for removing any hook, often without even touching the fish.

4. **Strive to Revive** – crankbaits with multiple sets of treble hooks can do more damage. Use extra care to do no more damage during the release. Check out Chapter Six for more helpful info on good catch-and-release practices.

No matter when or where you are fly fishing, remember the words of author Clare Vanderpool:

"Fly fishing is not about catching the fish. It is about enjoying the water, the breeze, the fish swimming all around. If you catch one, good. If you don't ... that is even better. That means you come out and get to try all over again.

AND NOW ... A LITTLE HELP FROM OUR FRIENDS

Presentation is everything. The fish know you're there before you get out of your car. As these fish spook so easily, this style of fishing makes you think harder and faster.

—Barry Carruth

Glasgow, Scotland

The Day it Rained Fish
By Jack Kerley © 2021

For years our family owned a cottage in Fairhope, Alabama, a quaint little town on the Eastern shore of Mobile Bay, and an easy drive to the Gulf. For our frequent trips to the beach, I packed both spinning and fly gear, hoping for the low waves and light wind allowing the latter.

One hot, bright morning, we aimed our headlights toward Fort Morgan beach, far less populated than the tourist-strewn venues of Gulf Shores and Orange Beach several miles east. We parked in a small public lot surrounded by subtropical flora. Palm fronds waved above. Gulls wheeled through the sky. Lizards skittered across the sand. We wrestled into our backpacks, grabbed chairs and fishing gear, and walked the long boardwalk to the beach, where a collection of beachgoers fanned out like a human delta. We turned sharply and left footprints in the sand until the nearest sunbathers were a football field away. Dunes and sea oats lay at our backs, and before us spread the sparkling Gulf of Mexico. The tide was incoming, waves small, a light offshore breeze further flattening the surf, the water clear enough to see your feet at a meter's depth; perfect fly-fishing conditions.

My wife unfolded her chair and retrieved a novel from her pack. I rigged my rod and reel, stuck spare flies into the patch on my cap, and strode into the water . . . where I stood for an hour, whipping loops above my head, casting and retrieving until my arms ached. I retreated to the shore.

"Nothing running?" my wife said, looking up from her book.

"It's the Dead Sea. What's for lunch?"

She rummaged through the cooler and produced zip-bagged pastramis and Swiss on rye. Instead of handing me my sandwich, she tossed it. I had to reach fast to make the save and gave her What's-with-that? eyebrows.

She smiled. "I wanted you to catch something."

I was working on a snappy retort when motion at the dune line caught my eye. I grinned and nodded.

"Old Tom's here."

Often while I fished lonely stretches of Alabama beach, we were joined by Great Blue Herons. They'd watch stoically until I nabbed a whiting and threw it their way or surrendered my remaining bait when leaving. Though always a different bird, we created a mythology in which all were manifestations of The One: Old Tom, Keeper of the Beach, a timeless entity with mystical powers.

"I'm not in Tom's good graces today," I said, affecting mock anger at the bird. "Dammit, Tom, do me some good and send fish."

Old Tom looked away and seemed to be stifling a yawn. My wife and I finished our lunches and she picked up her novel.

I said, "I should have brought something to read."

She pointed to the water. "Go catch me a fish."

It was the waning minutes of high tide, the waves undulating softly as slack tide approached. Knowing I wouldn't be spin-fishing (even without catching, I preferred fly fishing, the sign of a True Believer), I plucked my bait bag from the cooler—shrimp and cut mullet—and approached Old Tom.

He backpedaled and watched as I dumped the bait in the sand. As I withdrew, Old Tom approached. He stabbed at the feast with his beak, craning his head back as shrimp and mullet tumbled down the long neck. The sand was bare in 20 seconds. I turned away and knotted on one of two dozen Clousers tied in a hopeful bench session back in Kentucky.

"Round two," I said, stepping into the water.

The second session began like the first. I was starting another pointless cast when my wife shouted. I turned to see her knee-deep in the water, pointing downward.

"This is weird," she said. "Come see."

I splashed over, amazed to find a black basketball spinning just below the surface. Looking closer, it resolved into thousands of tiny black fish forming a swirling globe. The speed at which the fish moved and the delineated borders gave the impression of the first milliseconds of an ink drop hitting water.

"Another one." My wife pointed a dozen feet away. She splashed ahead. "And another." There were at least a half-dozen of the spooky dark globes slowly moving both down the beach and seaward.

Then . . . a fish as bright as a freshly minted quarter rocketed from the waves a hundred feet distant. Followed by another.

"Ladyfish," I said to my wife.

"Go get 'em."

I ran through the surf until close enough to launch a fly their direction. A third broke the water. A fourth. Two at a time. It was a school of ladyfish, bright and slender missiles; inedible, but fantastic fighters and jumpers. One took my fly, leaped, and threw the hook. I was retrieving line when another hit the fly.

I set the hook hard and brought it in, leaping and fighting. I was releasing it when I noticed smaller fish darting past, three or four inches, and suspected they were what the ladyfish were feeding on, and likely that the small fish had come to dine on the globes of tiny fish.

Next, hand-sized fish began popping from the water in all directions, some zinging inches above my head. They were shad spawned in Mobile Bay's estuaries and were entering their adult, saltwater phase. They were often mistaken for flying fish because of their leaping abilities.

I was dodging shad when white bombs started dropping. Sensing a smorgasbord, a cloud of gulls swirled above, dropping into the water with splashes like small depth charges, then, fish in beaks, returning to the air. They keened and wailed as another ladyfish hit my fly. I felt the rod dip and leaned back, setting the hook as another flight of shad shot from the water.

It began to rain fish.

They splashed the water around me, some whole, some only torn pieces, heads, tails—I saw an eye floating. I looked up and saw gulls having fierce mid-air battles over fish they'd caught. When a fish—or portion—lost to both birds in the battle, it fell to the water. I had a fish tugging my line, baitfish swirling at my waist, shad pocking the air around me like bullets, and fish raining from the sky.

It was an angler's glimpse of Paradise.

The phenomenon passed as suddenly as it had arrived. Motion ceased, fish scattered, disinterested gulls winged away, the shad tucked back beneath the surface. Somewhere in the frenzy, the ladyfish had escaped.

I reeled in my line knowing I had been touched by magic, blessed by the sea. There was no need to continue fishing. I returned to the shore, and we started to gather up our small camp. Old Tom was gone.

But as we were heading back, another heron stood at the dune line, framed by sea oats. Old Tom, of course. His eyes followed us as we walked past. I winked and shot a thumbs-up. I said, "Thanks, Tom."

I swear that big blue bird nodded and winked back, having done his good deed for the day.

Jack Kerley is the author of the Carson Ryder novels, with the thirteenth installment appearing in 2017. His short stories are "Almost There," published *in Southern Review,* **Spring**

2004, University of Louisiana Press, and "A Season of Moles," published in *Stories from the Blue Moon Café III: Anthology of Southern Writers*, McAdam/Cage, 2004, ed. Sonny Brewer. Kerley's books have been translated into ten languages and are published primarily now in England.

Best tip I ever heard is presentation will always out fish pattern.

—Rocky Locey
Ithaca, New York

This was the Gasper River
By Steve Champion © 2021

It runs about 25 miles in a northeastward direction into the Barren River near Bowling Green, Kentucky. More importantly, it drains the north-facing western edge of one of the limestone escarpments that produce the clear, limestone-filtered water for which Kentucky is known.

The headwaters of the Clear Fork begin at a spring near what is now Auburn, Kentucky. From this site, one can see 100 miles and the headwaters of a clearwater river in one frame. Certainly a rare sight, but also understandably sacred to the last three faith cultures that intentionally lived there since the beginning of time.

All my children have been in this river. I used it as therapy myself, many times, as I faced recent personal loss. It helped me re-establish a sense of permanence in my life. It is also widely known as one of the finest smallmouth bass streams in the Commonwealth.

Tecumseh's grandfathers taught their children that these streams, and their origins, were the sacred sources of all life. Even if my own faith tradition does not consider this particular place sacred, those who lived here did. I believe respecting that is the right thing to do— an important thing to do.

So, when I've seen trash on this river (or any waterway for that matter...), I felt that I owed it to Tecumseh's grandfathers to pick it up.

I wasn't, however, seeing what was ultimately going to kill the river. The youth camp downstream didn't see it. The church, 100 yards away, directly across the lawn, didn't see— or smell— anything.

Explained in quick, broad strokes, toxins from feedlots can cause a chain reaction that robs water of the 'dissolved oxygen' vital to life. Low dissolved oxygen is the *result*; toxins are the *cause*.

If conditions allow concentrations of the stream water to become toxic itself, rainfall can wash a "slug" of toxic water downstream bad enough to kill everything in the water.

Such an event occurred Memorial Day weekend 2019. Memorial Day.

Memorial Day.

I've taken this very personally. I'm truly heartsick. I don't know what to do first.

Here are some random numbers to consider to understand this event. The fish kill was documented covering 16 miles of river. A 21-inch smallmouth bass is likely ten years old. The feedlot was estimated at 200-plus head of livestock on only 2 acres. Feedlots haul feed in 18-wheelers that pour it in a long trough in one pass to minimize transportation costs. This feedlot used the clean river water for zero dollars. The operator appears to have received *tens of thousands of dollars* of Kentucky Conservation Resource Program subsidies to protect watersheds by using proper land management. He killed one river.

My river. Your river.

Consider that it will take 15 or 20 years to return life, as I was fortunate enough to experience it, to the Gasper River.

So, for me anyway, at 60 now, it looks as though I've joined Tecumseh's grandfathers in knowing this sacred, precious place only in my memory.

—Steve Champion
Owensboro, Kentucky

Authors' note – Stream restoration should top the list of priorities for all fishers. A portion of the proceeds from the sale of this book will be donated to Trout Unlimited for such purposes. We encourage all to properly care for the places we fish and be good stewards of our beloved sport.

1. Stop by your local fly shop. They will likely share some decent fishing holes with you, but do not expect the best holes. You will have to do more than buy six flies before they provide you with that information.
2. Hire a guide – it is now their job to get into fish.
3. Use an app on your phone. There are a bunch of them — some are decent and some are not.
4. Google best fishing holes in the area you are currently located.
5. Go by Sportsman Warehouse, Cabela's/Bass Pro Shops, Scheel's, etc., and see if they have a fishing board with local waters.

—Leo Poggione
Reno, Nevada

My Aunt Linda has invented a great way to keep people honest about how many fish they caught and released. All fish to the boat are given a name in alphabetical order of their entering the net—Arles, Barbara, Charley, and so forth. Sometimes for fun, we'll limit the names to a category like country singers or baseball players.

—Melissa Moody
Salem, Arkansas

Authors' Note – Rick has adopted this method, and being a huge fan of The Godfather movies, the sixth fish is always named Fredo. Before releasing the fish, he looks at the fish and recites Michael Corleone's words from Godfather II, "I know it was you, Fredo. You broke my heart." Apparently, not all fish have viewed The Godfather trilogy of movies.

5

You Can Make These Flies Yourself?

"O, sir, doubt not that Angling is an art; is it not an art to deceive a trout with an artificial fly?"

—Isaak Walton

We swore we would never do it.

Flies are not expensive. When we played golf, we lost golf balls. We never went home and made more in the garage. It is just too easy to run by the pro shop and buy a new dozen. So why, in the name of Isaak Walton, would we ever go and try to spin thread around a tiny hook when we both live five minutes from a sporting goods store selling all we need?

The entire concept of tying our own flies just seemed silly.

Then, one year on Father's Day, despite having heard both of us say we would never do it, Rick's wife bought him a fly-tying kit. It was a basic kit with a vice, a set of tools, and instructions on tying simple flies. The gift came with a gift card to the local sporting goods store to buy additional supplies and a reservation in their class for Fly Tying 101.

Rick never realized the gift was a setup from his wife – keeping him busy at his desk tying flies while allowing her to watch British baking shows in relative peace.

While Rick readily admits he has not yet "mastered" the art of tying flies, he spends his evenings peering through a lighted magnifying glass, trying to make simple twists of thread look like something tasty to a fish.

Admittedly there is a certain pride in catching a fish on your own handmade fly. It doubles the exhilaration. When asked by someone what you were using to catch a big fish, there is no more satisfying response than: "Oh, this little green nymph I tied myself."

The first key to tying is a good vise to hold your hook snugly in place while wrapping the thread. Secondly, if you are using cheaters to read, go ahead and buy a lighted magnifying glass. Finally, take a class in tying. Even during COVID closings, many stores continued offering classes online. There is also no absence of fly-tying books and videos on YouTube. Read them. Watch them. Be creative. And then practice saying, "Oh, it is just a little something I tied at my desk."

Rick's first flies were a mixture of colors and feathers he thought fish might find attractive. However, he soon realized the art of fly tying was much more than colors and feather tails. It was not like picking the right color of rooster tail lure. Instead, flies are constructed in line with the food chain in the water column mentioned earlier in the book: midge, nymph, emerger, dry fly, or streamer.

As we pointed out previously, there are 10 quintillion bugs in the world. Add multiple ties for each different hook size and two each if one gets hung up, and you will need a very large tackle box. If you add artificial flies imitating baitfish in both fresh- and saltwater, the number of choices multiplies dramatically.

So, tie for your water. But mostly, tie for fun.

As with fly fishing itself, we suggest you try tying flies first to see if you like it. Sign up for a class or sit down with a friend for some hands-on instruction. And the word "hands-on" is key. If you have arthritis, sight problems, or a lack of basic hand-eye coordination, you may find fly tying too frustrating to continue.

In this case, do not feel like a fly fishing failure. Stores and online sites offer every fly you will ever need to catch fish.

It is hard to describe tying without visuals, so we will not try. Instead, we would like to provide you with several good references for fly tying. In particular, we like www.flytyer.com, the Orvis Learning Center https://howtoflyfish.orvis.com/fly-tying-videos.

Below we will let some truly serious fly tiers take over.

AND NOW ... A LITTLE HELP FROM OUR FRIENDS

Authors' note – our friend Russ Galloway up in Maine not only makes his own equipment, but he has discovered a great way to add to the equipment adventure.

Getting into Fly Fishing on a Budget/Fixed Income

First off, let me state clearly that I am NOT an expert, nor do I profess to be. I am merely an older man that lives on a fixed income and has learned over my years to be frugal and do a lot of "re-purposing/economical shopping" to make the best of and get the most out of my favorite pastime.

Shopping for/Acquiring your Fly-Fishing Gear

If you are like me, you already have a curiosity/interest in fly fishing and have probably gotten "sticker shock" when you have looked at new fly rods. Yes, the big-box sporting goods stores have a better selection, advice, help, and information, but you're trying to do this on a budget/limited income, and this is where putting your boots on and putting in the groundwork comes in.

My favorite places to shop for fly-fishing gear/fly-tying materials:

1) Garage and yard sales/flea markets:

If you are driving around, yard/garage sale/flea marketing, always keep your eyes open for fishing rods. I have seen and bought fly rods/rod/reels at good prices. (*NOTE...ALL of my fly rods/reels are used except one, and I fish hard daily from spring thaw to winter's first ice for both food on the table and pleasure.) As with anything you buy used, check it over carefully, making sure it works properly etc. It helps to have an idea of what you are looking for/at as well.

2) Secondhand/Thrift stores/Vintage stores:

Some of these you already know; others are dotted along a major road in heavily traveled tourist areas and can offer a surprise or two as well. The more familiar ones, like Goodwill/Salvation Army, are also places to check as well. These might seem a long shot, and might be, but that one-time find makes it all worth it when you get it for a song.

If you wonder why I have not mentioned pawn shops as a shopping source, I will tell you why. They're in the business to make money.

They buy low and sell high, and 99 percent of the time have prices on quality rods that make shopping/buying on a budget almost impossible. Now, I am not disrespecting that option if you choose to go that route and buy a quality rod at their establishment. I just prefer other avenues and the extra "Benjamins" staying in my pocket for another day.

And this brings me to the last on my list of places to buy a fly rod/reel setup and be ready to fish in just a few minutes. Yes, it is Sam Walton's Walmart. Now, this said, they have two beginner sets (a 5-6 wt and 7-8wt) kits complete with rod, reel, and fly line all in one complete kit to be setup and go fishing. And yes, I have the 5-6wt setup (my only new rod/reel setup) and fish it and my other daily, as I have said before. In my opinion, it is a decent starter set if you are new to fly fishing and just learning as well as a backup rod/reel to have on hand just in case of "Murphy's Law" days out on the water...and trust me, everybody has them.

Tying Your Own Flies On A Budget

Tying your own fly does not have to be difficult nor expensive to stay on a tight budget if you have a few helpful hints to get started.

First off, flies do not have to be outlandish exact copies of professionally tied/store-bought flies to catch fish. Trust me, I tie my own, keeping to a few simple patterns and clones/off-shoots, and manage to catch fish and feed myself, too...even with the fly I tied with my two left feet.

Now I know you will ask, what about a vise and the tools to tie flies with?

What about a rotary vise versus a non-rotary? A portable vise versus one that stays at home on your designated tying table? How can I afford that one-hundred, two-hundred, or even three-hundred-dollar vise and stay on a "tight budget" and still make a decent quality fly that can/will catch fish?

Well, are you a bit handy and can you follow simple directions and want to make your own vise for less than fifteen dollars?

I can see a few of you going, "Yes...how?"

Now it is not pretty, nor is it smooth as a Swiss watch. Still, it is portable, it can be used as a rotary vise or stationary vise, and it will work to tie a simple fly for almost any species in the rivers and lakes, and it is cheap enough to make up and store at a cabin/camp in case of an emergency because the "tree fish" got all of yours.

Supplies For The Economy Vise Needed:
(1) 4-FOOT PIECE OF 1/2-INCH PVC PIPE
(2) T CONNECTORS TO FIT THE PVC PIPE
(2) HOSE CLAMPS
(1) 18-INCH PIECE OF 5/4 BOARD
(1) PAIR OF NEEDLE-NOSE VISE GRIPS
A HACKSAW AND A 3/8-INCH DRILL BIT, DRILL, AND A ROUND SANDING
BIT TO REAM OUT ONE OF THE T CONNECTORS FOR THE CROSS PIPE
4-6 RUBBER FEET FOR THE WOOD BASE (OPTIONAL)
DIRECTIONS FOR ASSEMBLY:

1. Drill a 3/8-inch hole in the center of the 5/4 board... ream it out slightly so that the PVC pipe will fit in snugly yet still be taken out with some effort.
2. Cut a piece of pipe to the height you want your vise to be/ to work with, remembering that you will be adding a T connector to it for the cross pipe.

3. Take the drill, put in the sanding bit, and ream out the T connector on BOTH sides inside so that the PVC pipe slides through snugly but can still be turned (like a rotary vise).
4. Once you have the T connector reamed and ready, with the pipe sliding through as needed, attach it to the top of the PVC post ... DO NOT GLUE IT ... you want this portable and able to be broken down for travel or to put away.
5. Slip the hose clamps onto the cross pipe (one from each side towards the connection...this controls the rotary part/the tension, etc., and keeps the cross pipe from sliding around, making a stable vise.
6. On one end of the cross pipe, add the second T connector, and on the opposite end, on one end of the pipe, cut a slit about an inch to an inch and a half long for the vise grip handle lock part to move in/out. Slip the screw end of the vise grips into the pipe and open them/close them so they lock and you will see why the slit is there.

And You Are Done...

And no, I cannot take total credit for this because I found it on the internet a few years ago and just tinkered/modified it to a more suitable model/usage. OK, you have your fly rod/reel all set up, you have your own "homemade vise," and you are ready to try to tie some flies.

Rule Number One — use the KISS method...Keep It Simple Silly. Fly-tying supplies like anything else can be or get expensive to get, especially on some of the fancier, more expensive, and intricate flies to tie. The choice is up to you what you decide to tie and what you use to built it.

Fly-Tying Supplies On A Budget/Repurposing The Scrounged And Found

Again, sticking to the "on a budget" principle, acquiring your fly-tying supplies for the simpler fly does not have to cost you a lot if you know where to do so. Here is a list of where and what I acquire for the basic fly that I tie:

1. 1) Goodwill (in the arts/crafts section: Goodwill is your friend. People donate to Goodwill, and quite often, you can find a blessing of basic supplies here including thread/s, feathers, yarn/s, Mylar/s, and "Easter egg basket grass" for making some simple and very effective flies.
2. 2) Dollar stores: These are a goldmine for supplies and all for just a dollar an item. Clearcoat nail polish "Hard as Nails" for a dollar a bottle, princess costume hairpieces in 4-5 different colors, bags and bags of Mylars including iridescent and several colors as well in the party section.

Car wash mitts in the automotive/car parts area that come in 2-3 different colors with the chartreuse and blaze orange being my personal favorites because they are chenille and used for mop fly and only require that you only remove the "fingers" from the mitt and end up with a bunch of long strands of fly tying "chenille" for many different flies. And these are just a few things that I know are there that work well for fly/fly-tying.

Scrounging/Repurposing Things Found

This is just a quick list of a few things you can "find" while out/about and can repurpose instead of leaving it lay or trashing it, along with why I choose to use it.

1. 1) Cloth-covered bungee cords…Got a broken one, one end missing or broken, or you find one laying on the ground? GOLD MINE…strip the cloth covering off it and look inside. What do you see? That's right, strands of rubber-like silly legs or a hula skirt just waiting to get re-used and be useful again…and it was free if you found it.
2. 2) Rope … cotton, sisal, poly ropes they all have a re-use purpose. Cotton/clothesline rope: fray out an end, and you have tail or dubbing materials that can be unwound and used as natural body materials. Poly ropes (yellow or water ski type ropes) can be stripped out by the strand and used in larger water/bass fly/predator flies. They are waterproof, durable, and sometimes difficult to work with, but well worth it when you get it right. Think of the

old "Fruit Stripe" Gum and picture it wrapped around a hook for the body and the colors you'd see.

Now, add to this all the tossed/lost phone charger cords, feathers, and dumped dead appliances you might find…stranded copper for a fly with a little effort. Feathers are always ways useful and the appliance power cord, too.

This is just a brief overview of how I have done it and how I continue to do so at every opportunity. I am not saying it is the right way nor the only way, but my methods have saved me dollars, helped repurpose some things, and even helped clean up the environment a bit too. And all for a sport and hobby that I love and enjoy.

—Russ Galloway
Westbrook, Maine

Author's Note from Rick – Since receiving this tip, I have started using the copper from old extension cords to make flies. Thanks, Russ.

My favorite fly is a pattern that is designed by myself. That pattern is called Gunna Fancy, but it's named after my mother. Her name is Guðrún, and Gunna is a nickname for that. The fly itself is fancy! So Gunna Fancy.

Here's the pattern and a photo of Gunna Fancy:

Hook: Kamasan B800 #8

Thread: UNI 8/0 black.

Tail: 50% red bucktail 50% black bucktail + 3-4 strands of pearl Flashabou

Butt: Red Glowbrite as a hotspot.

Body: UNI AXXXEL gold.

Beard: Red, yellow and light blue hen fibers.

Wing: Golden pheasant, and black squirrel on the top.

Head: Plastic eyes 2mm, sealed down with UV resins.

This fly hasn't been published here in Iceland, but I made a really good catch on it— so good that I will make a video about it on my YouTube channel.

Here's a link to my channel, Ívar's Fly Workshop: https://www.youtube.com/channel/UCtTAi5I0OBd6JsyYdVfvBbw

The channel is the largest and most viewed Icelandic fly tying channel. I make all my content both in Icelandic and English.

—Ívar Örn Hauksson
Iceland, Icelandic

6

Do I Catch And Release Them? Okay, Then How Do You Make Fish Tacos?

"I always catch and release and don't even eat trout. I think they are so majestic. I won't eat a trout even in a restaurant. They are beautiful and so much fun, and they give me such joy to catch them that it is my pleasure to take a picture with them, give them a kiss, sign a laminated autographed picture, and put it in their mouth."
—Henry Winkler, *I Never Met an Idiot on the River*

"To be able to serve and to eat a whole fish, especially a trout, is part of civilized dining. This applies particularly to the young, who should take to it as soon as they can handle knife and fork; this is a fine way for them to begin taking pride in themselves and their abilities."
—Julia Child

Well, those quotes clear one thing up. Henry Winkler never went to Julia Child's house for dinner.

But this chapter deals with one of the largest questions in the sport. Should fishers become sporting purists by catching and then releasing their watery prey? Or should they filet and eat the trout?

I suppose it all depends upon whose philosophy you choose to follow or, for that matter, how hungry you are on any given day.

Lee Wulff was an artist, pilot, fly fisher, author, filmmaker, outfitter, and conservationist. He believed any fish, especially trophy fish, are **"too valuable to be caught only once."** Wulff popularized the modern-day idea of catch and release fly fishing through his 1939 *Handbook for Freshwater Fishing*. **"The finest gift you can give to any fisher is to put a good fish back, and who knows if the fish that you caught isn't someone else's gift to you?"** said Wulff.

Comedian Mitch Hedberg cannot understand catch-and-release. **"They catch the fish and then let it go. They don't want to eat the fish; they just want to make it late for something,"** says an astonished Hedberg.

The noble act of catch-and-release is countered by the undeniable fact acknowledging properly prepared fresh trout are quite tasty.

And therein lies the fisher's dilemma. Release God's grandest of fish so it can live to fight another day? Or douse the darn thing in lemon and serve with a side of asparagus?

One day we were fishing the Cumberland River with two boats. We had caught some fish but nothing one would declare to be pushing the state record. Suddenly, a 23-inch brown trout rose to the top of the river and then rolled to its side. Wade's son, Colin DeHate, netted it out of the river and noticed something in its mouth. It was – and we are not making this up – a small carp. Essentially the brownie's eyes were bigger than its gills, and it choked to death in a lame attempt to swallow the smaller fish. The trout (sans the carp) was scrumptious.

A Fish Ate My Homework

We gave "CATCH-AND-RELEASE" advice for treble hooks a bit earlier in the book, but here are some classic catch-and-release concepts to consider.

Crimp the barb on the hook or use barbless hooks to lessen the impact on the fish. This is where the "tight lines" rule really rings true.

How long you fight a fish plays strongly into its survival capability. There is a balance between exhausting a fish past recovery and attempting to "horse it in" only to lose it. Bring the fish in as promptly as possible and use the rod (remember, the rod is your best tool) with side-to-side pressure to help you accomplish the task.

Use a net to land the fish. Wet your hands if you must touch the fish. Dry or gloved hands can damage the slime on the fish and make it vulnerable to bacteria and/or parasites. There are specific gloves limiting fish slime injury, so consider them for days when it is cold, or you still want your spouse to let you touch them after a day on the water. A good net on the fish is critical in not losing your fish and may ensure its ability to fight another day.

For the same reason as above, make every attempt to limit banging fish around on the rocks or shoreline if possible.

If you have really worn out a fish, gently cradle it naturally oriented below the surface and take advantage of the current. Many fishers recommend avoiding the back-and-forth action as it is not natural. Forward motion provides oxygenation through the gills. Be patient! When the fish is ready, it will swim away. If you do not believe you can successfully release the fish, then enjoy it for dinner,

For any method of fishing, use this age-old "rule of thumb" to consider what it is like to be "a fish out of water." Hold your breath when you take the fish out of the water. When you need to breathe again, guess what? So does the fish. If you need more time because you cannot get the hook loose or you want a picture, use the net to dip them in the water again while you are messing around with your darn smartphone and give them a sporting chance to revive.

Finally, just like everyone else, we want a photo of our prize catch. The authors of this book are no exception. We cannot count the number of times we have held up a fish for a shot to impress our Facebook friends. Just remember to get the photo quickly and release the fish. Both of us have been experimenting with taking photos and videos of fish while still in the water. It is harder to get a size perspective, but once you practice a bit, the images can be stunning – and even more impressive on your Facebook page.

Adventure and travel writer/author Jessica Maxwell adds this very important tip, **"What I had mastered was fly-fishing Rule No. 1: Remove all hooks from soft tissue underwater, where near-freezing temperatures anesthetize exposed nerve endings, and you can't hear your fellow anglers' hysterical laughter."**

More importantly, the experience of fishing gave rise to the Rick and Wade Rule in making life-and-death decisions for fish making it to the dinner table. If a fish swallows a hook and is bleeding out or otherwise cannot be revived, we declare "game-on" until we have enough meat for dinner. Otherwise, we catch and release.

I suspect this rule is not found in any book or guide, so feel free to appropriate it as your own.

What is the Science behind "Catch-and-Release"?

Every fisher you ask will probably say "catch-and-release" is a wonderful thing, and we should all practice it on any fish we do not intend to keep for a meal. We totally agree. The dilemma is our individual understanding of effective "catch-and-release" procedures does not mean we are all doing it properly to actually ensure the caught fish survives the release.

This is an important part of our journey into any type of fishing, and we are taking some time here on this issue to present some hard data to help all of us do "catch and release" better. Looking at *scientific data* (instead of well-intentioned opinions) will help us understand a percentage of fish caught and then released may die if there is unnecessary trauma to the caught fish from extended "airtime" or poorly understood handling practices.

Many of us have probably observed poor catch and release methods, but the impact on the fish begins earlier than we may believe and proceeds throughout the fighting and landing process. These impacts to the fish can be greatly reduced and/or completely rectified if we understand the benchmarks of what is happening to the fish.

In the interest of greater enlightenment on this subject, let's look at the following:
- For a fish, stress begins at the moment it is first hooked.
- The longer we fight a hooked fish, the more exhausted it becomes and (here's some of the science stuff) the greater the amount of lactic acid builds up in its body.
- When the fish gets landed and is huffing and puffing from the fight, we then:
    - Pull it out of the water from where the fish gets its oxygen, and
    - Pull it out of the water where the fish uses water to expel the toxicity of lactic acid buildup in its body.

The takeaway here is just when the fish needs the water the most we usually remove it from the water to get it unhooked and/or to get a "grip and grin" photo. I've heard several describe it this way: If a human runs up and down their stairs at home ten times, then sticks their head in a bathtub full of water, they would understand what the fish is experiencing. (By the way, do not do this, it is just an example).

So, let's add more science to these claims, affecting whether our attempt at catch and release is successful. In the June 1, 1992, *Canadian Journal of Fisheries and Aquatic Sciences,* authors

R. A. Ferguson and B. L. Tufts studied and published data on the *"Physiological Effects of Brief Air Exposure in Exhaustively Exercised Rainbow Trout (Oncorhynchus mykiss): Implications for 'Catch and Release' Fisheries."*

In our condensed and paraphrased summary of the research Ferguson and Tufts performed, they found:
1. Rainbow trout (*Oncorhynchus mykiss*) exposed to air for 60 seconds after exhaustive exercise initially had a much larger extracellular acidosis (lactic acid buildup) than trout which were only exercised. This is, of course, due to the inability of the trout to expel the acidosis through respiration in the water.
2. Survival of the exhausted rainbow trout dropped to 62 percent when exposed to air for 30 seconds and plummeted to only 28 percent in fish that were air exposed for 60 seconds.

These results show even limited air exposure is a significant factor in whether the fish survives, but what else may contribute to a fish not surviving?

In addition to the oxygen and lactic acid buildup, fish (particularly trout) have a protective slime layer preventing disease. When this slime is displaced by abrasion from any physical contact, including but not limited to a landing net, dragged across the shoreline, or handled with our DRY hands, it puts the fish at greater risk of disease even though it swam away after you released it.

Also, rough handling causes internal injury, resulting in the fish not surviving. Fish are meant to be properly oriented in the water where water flows through the gills. Undue pressure from our hands squeezing against internal organs causes injury. We cannot count the number of times we have seen fish dropped or squeezed during photographs. We cringe when we see people toss them like a baseball back into the water.

So, what are the takeaways from this?
- Consider using a at least the recommended weight of tackle to land the average fish expected. The weight/

strength of your tackle significantly determines the time it takes to successfully land the fish. Rod and reel combinations loaded with higher strength leader/tippet mean you can be aggressive in bringing the fish to the net, so a shorter fight is better for the fish. Of course, there may be times when you hook fish, and they were overwhelmed by your tackle, but if you regularly are using lightweight tackle for the commonly caught fish in your area, perhaps it is time to consider an upgrade in tackle weight.

- Consider never touching the fish. A quick pop of the hook from the mouth of a fish alongside the boat or shore results in less handling and stress time.
- Consider a soft rubberized net to reduce slime removal for those fish that cannot be unhooked easily. If possible, avoid touching the fish.
- As silly as it sounds, prepare for a photo opportunity ahead of time. Have your phone/camera ready, and designate who the photographer is for the angler(s) ahead of time if you can. This will *greatly* shorten the airtime for the fish. Just as a thought, the next time you catch a fish, try to hold your breath from when you take it out of the water until you put it back. You get the idea.
- If a photo must be taken, hold the fish *securely* around the base of the tail with one hand and *cup* the mass of the fish's body with the other hand (do not squeeze). This method is less traumatic for the fish's internal organs to be securely but properly held than to be dropped in the boat or on the shore because you were not holding it securely. Better yet, take a picture of the fish in the water or in the net while still in the water! As we mentioned earlier, there are many nets with size markings in inches (with zero at the bottom of the net), so you can quickly gauge the size without excess handling.

- Consider the following when returning the fish to the water:
  - Place it headfirst and hold it in normal orientation in the water. If it is lethargic, help revive the fish by cupping the body and tail so the fish is oriented properly, *and facing into the current.*
  - Avoid a back-and-forth motion as fish generally do not have a reverse.
  - When the fish attempts to swim away, turn it loose!

In a March 21, 2019, article by Eric Barker in *The Lewiston Tribune*, he writes that "**a coalition of conservation groups that sought to shut down or alter Idaho's steelhead fishery. The groups said that the return of wild steelhead to Idaho Rivers this fall was so low that all fishing should be stopped or that practices such as fishing from boats, using bait, and lifting wild fish from the water prior to release should be banned.**"

In an October 11, 2018, article by Eric Hayes in the *Pittsburgh Post-Gazette*, the "catch-and-release" method was also criticized in a 2018 study by the California University at Riverside of fishing mortality published in *The Journal of Experimental Biology*. The study cited concern over mouth injuries suffered by fish caused by hook removal after catch-and-release hampering the ability of suction-feeding fish, such as bass, salmon, trout, and perch, to capture prey.

Kristopher M. Kuhn, chief of the Fisheries Management Division at the state (Pennsylvania) Fish and Boat Commission, said, "Sport fisheries are managed at the population level, not with the intent to save the life of an individual fish. Incidental loss of an accidentally caught and poorly released fish does not threaten the population."

As a result of these concerns, several studies were conducted to determine actual catch and release practices being used by anglers.

- In a 2016 article by James Lamansky Jr. and Kevin Meyer published in *the North American Journal of Fisheries Management*, anglers exposing trout to air averaged about 29 seconds. Only 4 percent of anglers held trout out

of water for more than 60 seconds. Fight time averaged about 53 seconds. "**When photos were taken, it extended the time out of water by an average of 10 to 20 seconds.**"
- Also in 2016 and 2017, Idaho Fish and Game biologists Luciano Chiaramonte, Don Whitney, and Joshua McCormick found when anglers took a photo of a steelhead, they held it out of the water about 1.7 times longer than those who didn't take photos. Fight time averaged 130 seconds, and "*those using fly fishing gear took an average of 1.54 times longer to land fish than those using traditional gear.*"

In saltwater fisheries, the Florida Fish and Wildlife Conservation Commission states, "If the fish is hooked deep in the throat or gut, research shows **it is best to cut the leader at the hook and leave the hook in the fish**... Fish are capable of rejecting, expelling, or encapsulating hooks." Likewise, Bartholomew and Bohnsack at the National Marine Fisheries Service revealed six studies demonstrating *higher mortality with natural bait than with artificial lures and flies* while five other studies found no difference. "**Natural bait led to higher mortalities than artificial bait for cutthroat trout** (Pauley and Thomas, 1993) and rainbow trout (Schisler and Bergersen, 1996)." In the same research, "**No differences in mortality from single vs. treble hooks** were reported for non-anadromous trout (Taylor and White, 1992), cutthroat trout (Pauley and Thomas, 1993) and common snook (Taylor et al., 2001)."

Finally, we firmly believe catch and release is a successful practice to protect our fisheries. We always hope any fish we do catch and release survives. However, after an honest effort to catch and release properly, if the fish does not survive, it is usually not wasted. Instead, it was taken care of by the circle of life. We commonly see blue herons, osprey, eagles, pelicans, and an occasional coyote, cleaning up casualties near fishing locations. But ultimately, we all need to improve and practice better catch and release.

And while, even considering our one exception to our catch and release, there is no denying freshly caught trout is absolutely delicious. Trout can be prepared in any number of ways: pan-fried, oven-baked, smoked, grilled, wood-baked … the possibilities are literally endless. Do a quick search on the internet, and you will find more ways to prepare trout than there are ways to catch them.

Well then, What's for Dinner?
If you do take some trout back for dinner and want to get a little fancier with your catch, here are some recipes to consider.

Skillet-Fried Trout
Bread the filet very lightly in flour and sear in butter using an iron skillet. Season with lemon juice. Some people like to add parsley or a touch of salt or pepper, but others (like us) really like the fresh unseasoned taste. Fry until the meat is flaky. This recipe is particularly easy for a campsite meal.

Baked
The best way to bake trout is in tin foil. Baking dries out the meat so quickly it needs to be surrounded by the juices offering it flavor. From a simple butter and lemon concoction to a spice-laced mixture, baking provides a fine moist filet. The taste with this method is usually improved by garlic cloves or herbs, like rosemary or thyme.

Wood-Baked
For trout baked on a wood plank, you generally need a thicker filet. The filet needs to be cured with a dry rub like what is discussed in the smoked section. Try maple syrup instead of sugar or honey. It enhances the wood taste. Placing the cured filet on a soaked wood board and cooking in a wood-fire oven (or barbeque grill) allows the filet to take on the taste of the rub as well as the wood.

Grilled

Easy. Toss them on the grill and cook until flaky. The big question becomes full fish or fillets. Some people just do not want their meal to be looking at them. However, grilled fish are best full. Add lemon and salt to flavor.

Smoked

The key to smoking trout, or any fish for that matter, is the brine. Either wet or dry, the brine is what determines the unique taste. Be prepared to experiment until you get the taste you crave. A wet brine starts with a liquid ranging from water to beer. Salt needs to be added to keep the meat moist, and sugar (or honey) is added for taste. The spices you add depend on the flavors you are trying to elicit. Keep the fish in the brine for at least 8 hours but let it dry before smoking. Choosing the wood for the smoker is important for flavor. Fruit tree wood gives a milder taste, while others like oak and mesquite give the fish a stronger flavor. Cooking time and temperature usually depends on your smoker but expect about 3 hours of cook time. Many people prefer smaller trout when smoking with a wet brine.

Cooking with the dry brine is not much different from cooking with wet brine, except you rub it into the meat instead of marinating it. There are no liquids involved, and brown sugar replaces the honey. Dry brine requires you to cook a little slower, so expect about 5 hours in the smoker.

In both wet and dry brine, skin on or skin off is a personal choice.

More Options

Pinterest is a great place to go for trout recipes of all kinds. There is a Facebook page dedicated to preparing trout. https://www.facebook.com/Rainbowtroutrecipes/

Pairing Wine

You can find quite a discussion online about what wine pairs best with trout or other fish. Those with the palate for a drink enhancing the taste of food, will spend hours arguing the point.

The authors will settle this once and for all – bourbon. Rick prefers Michter's, and Wade likes Makers Mark. And, if it was a particularly noble fish that put up a good fight, we will break out the Macallan's Single Malt Highland Scotch. There. It is settled. You are welcome.

AND NOW ... A LITTLE HELP FROM OUR FRIENDS

TIM FARMER'S SMOKE TROUT ©

8 cups water
4 trout
¾ cup kosher salt
½ cup brown sugar
½ cup white sugar
2 tablespoons garlic powder
2 tablespoons black pepper
1 tablespoon lemon pepper
½ cup soy sauce

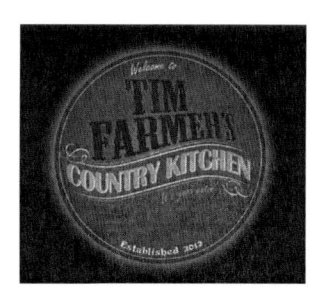

Mix together ingredients and brine trout for 6-8 hours. Remove from brine and let dry (can even use a fan) for 30-45 minutes. Oil smoker racks and place fish on top.

Let smoke for 2-4 hours at 200-225 degrees, or until internal temperature of fish reaches 160 degrees for 30 minutes.

—Tim Farmer

Tim Farmer is the host of Tim Farmer's Country Kitchen. For additional recipes, cookbooks and showtimes, please go to www.timfarmerscountrykitchen.com.

I never used to understand that release part of fishing until I started fly fishing. I now fish almost exclusively with barbless hooks, and I release everything. Once you have that 20-inch rainbow take off on you upriver, and then you fight him all the way back to the net over 5-10 minutes, you will understand why. There is a very good chance that someone before you also caught that same fish, and they released it back into the river to grow and someday be caught again.

—Leo Poggione
Reno, Nevada

Authors' Note - We asked our friend (and master chef) John Mocker for his favorite trout recipe, and he gave us this variation on a Julia Child dish.

Pan-Fried Whole Trout

As mentioned in Child's quote introducing this chapter, I like to use whole trout and pan fry them in butter. The trout should be dried and the inside cavities seasoned with salt and pepper. Place a light coating of flour over the fish. This will help the fish keep it shape while it crisps. Add the flour just before you are about to sauté it. If you wait too long, the flour gets wet and spoils the presentation.

Sauté each fish in clarified butter for about 5 minutes on each side. You can tell they are done when you can just separate the flesh from the bone at the ridge of the spine. Make sure there is no rosy color near the bone in the cavities.

When serving, sprinkle with hot browned butter.

—John Mocker
Union, Kentucky

Campfire Trout

Every part of this is the way it ought to be. From the experience of gurgling cold water flowing around your legs, to the colors of the stream, the fight, the conquest, the smell of smoke, and the ultimate aroma, flavor, and satisfaction of the meal.

Start with:
4 small trout cleaned and butterflied
Folding grilling basket
Salt and Pepper
1 lemon, thinly sliced
Your favorite herbs, such as rosemary, fennel, thyme
4 pieces thin-sliced bacon

Build your campfire and drag the cooking coals into a pile away from the main fire. Place your grate above them. Lay the butterflied trout in the grilling basket opened up flesh side up.

Sprinkle with salt and pepper, placing lemon slices on the flesh side of the trout and the herbs on both sides. Fold the fish closed and wrap each with a slice of bacon.

Sprinkle more salt on the outside of each trout. Close the grilling basket. Place the grilling basket over the coals on the grate.

Cook the trout, turning the basket over occasionally until the trout are browned and the bacon is done. Remove from the heat when flesh flakes with a fork.

Serve on a plate with a fresh lemon slice and rosemary sprig.
Wine suggestion: Rosé

—Marcus Carey
Owenton, Kentucky

Marcus Carey is an avid outdoorsman and host of "The Marcus Carey Perspective" podcast.

Rainbow Trout Tacos
By Colin DeHate

Ingredients:
1 ½ pounds Rainbow trout
1 ½ T olive oil
1 red pepper
1 yellow pepper
1/2 red onion
½ tsp. Cumin
½ tsp. Garlic powder
½ tsp Cayenne pepper
1 tsp Salt
¼ tsp Black pepper
1 Lime
Fresh Cilantro
Romaine lettuce
1 avocado
Shredded Colby Jack Cheese
Sour cream
8 large tortillas

Directions:
Combine cumin, garlic powder, cayenne pepper, salt, and black pepper in a small bowl with a lid. Shake until spices mix thoroughly.

Remove seeds from peppers. Dice peppers and onions and place in skillet. Drizzle olive oil over peppers and onion and sauté on medium heat. Salt and pepper to taste while cooking them down until onion and peppers are almost browned.

Cut trout fillets into thin strips and cube. Set aside.

Cut avocado into strips, and shred lettuce. Set aside.

Finely dice a small handful cilantro and cut lime in half. Set aside.

Spread spice mix over trout and dry rub into the cubed fish. Place tortillas in 250° F oven to warm.

Once peppers and onions are cooked down, add seasoned trout and cook 3-4 minutes over medium heat until fish is no longer translucent. Remove from heat.

Add cilantro and drizzle lime juice over cooked trout and pepper/onion mixture.

Remove tortillas from the oven and dress them with lettuce, sour cream, cheese, and fish mixture. Enjoy!

7

CAN WE ALL JUST FLY FISH TOGETHER?

"The great thing is to last and get your work done and see and hear and learn and understand; and write when there is something that you know; and not before; and not too damned much after..."
—Ernest Hemingway

Chapter introduction by Wade DeHate

If you and your partner like making memories together, you are not alone. I love it when my wife goes fishing with me for a couple of selfish reasons. First, she is (predominately) unselfish. Notice I said "predominately" as her weakness is to eat all the best buffalo wings on the plate first. The second and actual reason is even though she does not always fish as much as me, she does not want to be a liability when we are fishing together. If she sees I'm having a good time, she will never suggest we leave. So, if you want to consider an angling partnership, consider the following recommendations:

- Get your significant other trained and equipped, so they are self-sufficient; otherwise, they are sure to feel like baggage. If you are patient (and they are receptive), train and equip your partner yourself. Otherwise, take a class offered by a fly-fishing shop.
- Consider where they want to go, not necessarily your favorite spot. There are some nice places within a reasonable drive, but they may not be your personal favorite. An alternative is to take a trip for couple of days or get a fishing guide, so there is no work involved (think: going out to eat and not having to cook dinner).
- Create a successful environment. Focus on your spouse catching fish as discreetly as possible to get (and keep) them enthused. While fishing, it is very easy to get back in your own "disconnect" zone and ignore others. Pay attention to your spouse.
- A picnic is a good thing. Plan for a comfortable day on the water. Going out there to rough it probably will not make for a fun day.
- Be prepared to get out-fished.

By the Way – She Started It

Fly fisherwomen then and now…

If you are not already aware, the sport of angling needs to embrace the fact the ladies have the honor of first documenting the sport of fishing, especially the use of artificial lures, flies, or some type of faux bait. Dame Juliana Berners authored *"A Treatyse of Fysshynge wyth an Angle"* in the age-old year of 1496.

Yes, friends, this makes her work the earliest known documentation of the sport of fishing. Her *"Treatyse"* predates Izaak Walton's *"The Compleat Angler"* by a solid 150 years.

According to Brittanica.com, "Berners was a noblewoman and prioress of the Sopwell Nunnery near St. Albans, England during the late fifteenth century. (Oxford defines a "prioress" as "a woman who is head of a house of certain orders of nuns"). Various accounts of the history of fishing literature describe her as a woman of keen intellect and an accomplished practitioner and avid devotee of outdoor sports, including angling and hunting."

Dame Juliana covers a broad scope of the intricacies of fishing, such as the type of rods and the line used as well as the selection of lures and flies, and goes as far as the best choices by season. On top of that, she espouses thinking ahead to preserving fishing resources and proper etiquette when fishing with others. Brittanica.com goes on to confirm, **"These concepts would not come to be commonly accepted and advocated in the angling world until 400 years after the publication of the Treatyse, yet today they embody the ethical bedrock of sport fishing."**

It is worth the read, but Old English makes it a bit difficult at times.

Jim and Janine Young are the exception to the fly-fishing spouse rules. They are both guides and tenacious fly fishers. Jim once told me Janine never hears the words, "One more cast." Here is some prolific advice from a couple of great fishers who have traveled the world in search of the perfect cast.

Couples Fishing
By Jim and Janine Young © 2021

The Tasman River lies in the valley below Mount Cook on New Zealand's South Island. For us, it was one of the most beautiful locations we had ever fished. Blessed with a month on the South Island for fishing, we fell in love with the country, the people, the food, and especially the rivers.

Fly fishing the Tasman for trout is a very different experience in February from our late summer mountain streams in Colorado. Most of the fish, especially the smaller ones, have returned to Lake Pukaki downstream. Remaining to inhabit the larger pools are the largest of these truly wild fish.

Fishing for them feels like a combination of big game hunting and fly fishing. Since fish numbers are low in the river at this time, you stalk fish over long distances, walking the shore while searching the water for the tell-tale shadows of large fish. A skill that takes some developing on larger water and one you hope your guide possesses to a high level.

Once you have spotted a target, the stalking begins as one of you and your guide work to get into position for the few casts these extremely wary fish will tolerate before bolting downstream and out of sight. The other fisher gets to take a break from the walking and, you hope, get into position for a few photos should the fight and fish turn out to be impressive.

Upon approach, the fish was visible just below a drop-off into a large deep pool, rising to feast on passing nymphs. A few delicate casts, and the fight was on. Multiple runs upstream and then full-speed downstream brought loud screams from reels, guides, and fishers.

Resting on the opposite shore, your partner was in a great location to capture scenes of the fight. Multiple jumps allowed even a mediocre photographer to capture fish midair as the battle continued. The contest ended with a careful netting, respectful release, and smiles worthy of a million-dollar lottery win. Even I, Jim, having played only the part of photographer in these memorable few moments, was smiling.

You might be tempted, as you begin reading this chapter, to assume my wife Janine and I are some sort of special couple. You know, grew up on the shores of some famous fly fishing river, Introduced to fly fishing by our wise sage grandfathers, etc. Spent our childhoods traveling the world and fishing its greatest fly fishing destinations. You should drop any such preconceptions from your mind immediately.

Jim's Story

I (Jim) was introduced to fly fishing at a young age and did grow up on the banks of a well-known river. But that river was the Ohio where I never saw a fly rod, and the sage who introduced me to fly fishing wasn't a grandparent, but rather Curt Gowdy. While my friends were outside my backdoor playing the sport of the season, I would collect my Oreo cookies and 8-ounce

Coke in front of our black and white TV each Sunday afternoon, preparing to embark on a great adventure with "The American Sportsman."

In that brief hour, I would travel to lands I had never heard of, hunt big game animals I'd only seen in zoos, and watch the melodic casting of the hosts as they fly fished. It was always an amazing adventure for my young spirit, but each week I hoped for segments featuring fly fishing or waterfowl hunting.

Well, TV shows get canceled, and a young boy's dreams are overcome by fast cars, playing sports, and girls. And while there were many attributes I desired in girls and a future wife, the idea of a fly fishing partner never made the list.

It wouldn't be until my late 20s, on a trip for a training conference in Vail that I would see my first honest-to-goodness, live fly fisher. As we rafted the somewhat tranquil "whitewater" that day, I was suddenly mesmerized by a lone fisher ahead, artfully casting into a soft riffle. I did not see him catch a fish. He barely even recognized our existence, somehow intently connected to the river and his fly in a way I hadn't noticed in my extensive training by observing Coach Gowdy. Everything seemed to stop as we passed, feeling like I had violated the sanctity of a priest in prayer.

Over the next several years, my interest returned, but the opportunities for fly fishing were severely limited in Northern Kentucky in the '80s. So, my outdoor passions were more expressed in spin casting for bass, archery hunting for deer, and most avidly hunting waterfowl—another interest planted by those early "American Sportsman" shows. But women were few and far between on "The American Sportsman," so Janine was never invited along on these days afield.

It would be nearly another decade, after I began a career working as a fundraiser for an international conservation organization, that a business gathering in Montana would break open what is today my greatest outdoor passion. No fish were caught that day either, but that first experience of floating the Missouri River in a drift boat, borrowed fly rod in hand, was one of those moments after which life is never the same.

Janine's Story

I (Janine) was along on the trip to Montana but not invited into the boat that day. My journey, while geographically close, was much different.

Growing up in an angry, tense home with outdoor adventures miles away, I found peace and solitude going swimming at our local pool. Water was soothing, fun, and pressure-free, so I took every opportunity to enjoy the solace it provided. Water in the form of a pool, a pond, lake, river, or small stream has been a theme in my life. Peace comes in wild places.

So, the outdoors has always been a haven for me including when it came to fishing. My first fishing experiences were on my aunt's small farm pond with my cousins, mom and grandmother. Laughter and fun were the day's agenda, as well as the pursuit of bluegill, and then excitement if a fish were to tug a line. When I was 8 or 9, I walked up to my grandmother and told her I could not put this worm on my hook. The worm was dangling and squirming between two fingers an arm's length away. She very calmly took the worm, spread it out on the firm clay near the water's edge of the pond, and said, "Stretch the worm out straight like this and smack him a few times, and then he won't squirm so much." It was true! I was able to put the worm on the hook myself. A small kind of accomplishment for a child, but I knew I could do it. That encouragement brought confidence in a wild place that was unpredictable and fun.

Looking back at the times around the pond, I realize I experienced a community of women: my mom, aunty, and grandmother with myself, my sister, and three cousins — two of whom are beautiful women today. None of the men of the family were around the pond with us. I find that very curious that my introduction to the outdoors and fishing was through women. It was beautiful.

With so much appeal and connection to the outdoors for me, a place to be my own person away from family demands, I, like Jim, was drawn to the Saturday morning fishing shows with Curt Gowdy and so many other accomplished anglers. I dreamed of being out on the water in a boat or standing in a stream beside them, joining in on the adventure – and catching fish. I did not understand that I may not be welcome in the boat because I was a female. At that time, I did not see women fishing, at least not fly fishing. I now know there is a long history of women fly fishing, including a woman, a nun, Dame Juliana Berners, who in 1496 wrote, "A Treatise of Fishing with an Angle." The first book on the subject of fishing printed in England. I had no idea at the age of 9 that I would be joining a long history of women fly fishing.

I was on the trip to Montana Jim mentioned. I was in my early 30s, and it was my first trip to the west. I was awestruck looking at the land, mountains, and rivers that flowed in meandering ribbons cutting a path through sagebrush and willows. There was nothing like this in the landscape I come from. Wide-open spaces and big sky. The air was fresh and cool while the sun was shining so brightly that I had difficulty looking out across the fields. I felt at home in this wild, wide-open space. Then I saw it, a boat calmly floating down this beautiful river of the clearest water I'd ever seen. There was one person in the front of the boat and one person in the back of the boat, both casting a fly line as poetry on the wind with unique precision. I was captivated, and the dreams I had of being in the boat while watching a small black and white TV came back to me.

Our time in Montana did more than break out my (Jim's) passion for fly fishing. It birthed a love for the West and the Rockies between both of us. We spent much of that trip talking about how we felt at home in the mountains and how we had to return, maybe even move here.

Before our time in Montana, work had relocated us to East Tennessee. So, I took advantage of that borrowed fly rod and began trying to learn my way around the rivers of the Smokies. It was tough going, and I suspect, not very pretty to those who encountered me on the river. But I scratched out a few fish while Janine watched from the shore, usually trying to keep our two young boys from throwing rocks into the water Dad was fishing.

Soon we got our dream of a move west, just overshot a bit, landing near Sacramento. We did some fishing, caught our first salmon on spinning rods, and continued to fall in love with the mountains, the West, and trout. I didn't own a fly rod, but apparently was so frequently talking about fly fishing and always asking questions of anyone who did fly fish, that 18 months later, when I took an opportunity to relocate to Denver, my team gave me a custom fly rod as a going-away gift and said, "Quit talking about it and go do it!"

I arrived in Denver and immediately signed up for an Orvis Introduction to Fly Fishing course and was off and running. With abundant access to great rivers within a few hours' drive, I began spending every available moment on the river. I began reading, asking questions of people I would meet, going to fly fishing shows, hanging out in fly shops, looking for all the direction I could get. I usually dragged Janine and the boys along with me to reduce the guilt of spending our family free time pursuing my hobby alone. Thankfully they were generally good sports about it.

While tagging along to various locations, I (Janine) was also listening. Mind you, I usually was watching both of our boys at the time, so I just got tidbits, but I was also fascinated by this sport that you get to be in the stream, walk upstream, and fish to your heart's content. In other words, you didn't have to sit on the shore and wait for a potential fish to take whatever was on the end of your line, watching for the bobber to move. Honestly, that was so boring to me. No, I got to look for fish, then cast to fish in the hope that a fish would eat the beautiful, tiny fly on the end of what I now know as tippet. Tippet so fine and delicate that it is hard to see and yet strong enough to catch beautiful trout.

On one of my tag-along trips, we went to Yellowstone National Park over our boys' fall break. While there, I sat and watched Jim standing in the middle of the Yellowstone River casting to elusive trout. I knew then that I didn't want just to watch; I wanted to do this beautiful, challenging thing called fly fishing. So, when we got back home, about eight hours away from Yellowstone National Park, I picked up my first fly rod and reel. The motion of the cast and the feel of the fly rod was definitely challenging, and while I wanted to be proficient right away, I knew I couldn't, which brought me a great deal of frustration.

Like anything new, every experience is practice. I eventually got a pair of men's neoprene waders and wading boots; at the time, women's gear was not available. Before long, I was standing out in a stream casting and looking for fish. Honestly, the catching came later.

With both of us and the boys having been infected with this addiction, our lives took on a great focus toward rivers.

As Jim and I were still learning and practicing fly fishing, we were also trying to figure out how this would work for us as a couple. You see, Jim is a coach by nature, so he would naturally coach me in my practice out in the stream, sometimes the coaching was welcome and sometimes not. I remember him being yards away, yelling for me to keep my elbow down or not go so far back in my backcast. Then inevitably, I would get a knot in the line or get the fly caught in the willows behind me. Oh, those early days were challenging. Yet being in the stream, hearing the water running over boulders, feeling the wind on my face, and seeing it blowing through the trees made my experience. I wanted to be out on the water as often as I could be.

We went back to Yellowstone National Park the following year, again over our boys' fall break. By this time, both of our boys were fly fishing, too. Ethan, our youngest, was 8 and was a natural at casting even though he had many knots I had to untangle. We had such fun. I remember it was on this trip when I knew I could actually get proficient at fly fishing. Jim, Ethan and I were fishing in the Lamar Valley on a beautiful small meandering stream with gin-clear water. Jim was way upstream of Ethan and me but had been standing in the same spot for a long time. We finally worked our way up to him. It turned out that there were some beautiful rainbows tucked under a boulder that was jutting out into the stream. Occasionally, a fish would come to the top as if in slow motion and take a bug off the surface of the water and slowly go back down to its hiding place. Ethan and I took a break sitting down back from the shore and watched Jim try to get one of the elusive fish to take his fly. While I was sitting, I took the opportunity to put on a new leader, tippet, and picked a Royal Wulff to tie on. Jim was frustrated and tired by this point, so I said,

"Why don't I give it a try?" Jim said OK, so he stepped aside and let me get in a position to cast. I got my line and fly ready and put out a beautiful cast landing the fly just upstream from where the fish were rising. As if in slow motion, my fly gently floats down to the boulder. I see a fish come cautiously to the surface and then take my fly off the surface. It was so beautiful and perfect! I really can do this! Then came Jim's response, "You dirtbag! What did you put on?" While I confess, I was shocked, I was also delighted and laughed. I knew at this point our relationship on trout streams would have a new element…competition!

We had some tough times in those early years because of this thing called competition, but we both hung in and learned what some boundaries needed to be for both of us. Now, we both want to catch fish, that's true, but we can also learn from each other and respect the abilities of the other. We knew we didn't want competition to ruin the thing we loved to do together and ruin the experience of being on a trout stream, seeing fish rise, watching the swallows darting over the surface of the river to catch a bug emerging from the water's surface, hearing the wind as loud as an airplane in the trees, and enjoying a glass of wine on the tailgate of the truck by the river at the end of a beautiful day fly fishing.

So, when everyday life gets stagnant and routine, we escape to the solace of rivers, mountains, and trees–a place that is timeless and quiet, when we become aware of what is being held so tightly within and can let it go. One of my favorite poems is by Wendell Berry, "The Peace of Wild Things." We can go to a trout stream and find our humanity. When people ask me if I eat any of the fish I catch, I always say, "I don't eat my therapist." Therapy for me is standing in the middle of a trout stream, and if a fish takes my fly, I bring him to my net and let him go. Swimming out of my hand, I also can release all the things I catch in everyday life.

The opportunities for fishing together are endless. We've fished almost all across the United States. We have fished in Canada, Alaska, and Mexico and just recently, we spent a month together fishing the South Island of New Zealand.

Tips for Husbands from Jim

No matter how well it may fit, or how appropriate you think your old gear may be for your wife, do not use your hand-me-down equipment for her. Today, there is an ever-increasing amount of equipment designed specifically for women. When we started, there was not much ladies' gear. What existed was often referred to by experienced women fly fishers as the result of the "pink it and shrink it" approach. It mainly was just men's gear colored or trimmed in pink and made in a few sizes smaller. If you truly want her to enjoy and thrive, encourage her to buy good equipment that she likes.

Do not think you are going to teach her how to fish. Since Janine will certainly read this, I'll skip the reasons for this one since we might disagree. But do not take the approach that you will learn and pass it on to her. If you have been fishing longer, hire a guide and let the guide focus on your wife. You will have more time to fish instead of helping her, and she will learn more from the guide (who is not her husband even if you are a guide) than she will from you. She will ask more questions and be more responsive to his repeated directions. He is teaching; you are just nagging and being critical.

One great opportunity for both of you to grow is to attend fly fishing shows. Most of them have lots of teaching sessions. You can attend the ones together where you both have interests. But you can also split up and attend some by yourself, focusing on those parts of your skills/knowledge you would like to improve. These are also great places to check out possible fishing destinations together.

Do not expect her to fish like you do. Do you and all the guys you have fished with in the past fish the same way, prefer the same rod and waders, start with the same rig, etc.? Since I started before Janine, I usually expected her to do things the way I did. But it did not take long for her to develop her own ideas and approach. For example, I would always cover lots of ground. We hit the river and agreed to move upstream, hopping each other as we moved. Soon, I would be a quarter-mile upriver and she would still be in the same hole she began in. So, I would return downriver, provide a lengthy explanation about how you can't expect to catch anything here after all those casts. You've spooked every fish in this hole and need to move.

This explanation was usually followed immediately by a smirk or rolling of the eyes. Then soon followed by the sound of a fly line tightening up on a fish. She will learn and grow differently and frankly it will not be long until you ask "how did you do that" or "what fly are you using" more than you will want to admit to your friends.

Get over it; she will cast better than you. While it is not a universal truth, it is about as close as things come in the fly fishing world. Women learn to cast a fly rod quicker and better than men. We men suffer from the "get a bigger hammer" philosophy of life, and this carries over to our casting. Men assume that if the cast was not what I wanted it to be, I must do it harder and faster. This is almost always the wrong answer. When I teach casting to a guy, my most often first exchange sounds like this.

Great, now do it again half that hard.

Good, that was better, now half again that hard.

Better, now one more time, half that hard.

After this adjustment, most men are only casting twice as hard as they need to.

Fly-casting is about timing, technique, and rhythm. Women get this so much easier than men generally. But again, do not try to teach this to your wife either. Let a friend teach her or get her instruction, if available, from a female instructor. Your only hope for out-casting her will likely be distance. But practicing together can be a lot of fun. And occasionally check your stroke, pace, and amount of energy put into your cast against hers. You might learn something.

Even on the river, she is still your wife, not just another fishing buddy. Once your wife develops some skills, catches more and bigger fish, and is clearly as good of an angler as you, you might feel the temptation to start treating her like a buddy. DON'T. She is still your wife and a lady. There will be things where she can use your help—crossing the river, for example. I'm 6'5", Janine is 5'6". What is waist-high water to me is chest-high water to her. So, we often go across deep or fast water together.

Your partner probably did not grow up playing in the woods, mud, streams etc., that you did, so help find your way to the river, etc. Basically, this is something you are doing together, even though you may be slightly separated. The goal is to finish a great day of fishing in a way that makes for a great drive home sharing the memories of the day—not working through your steps for resolving conflict.

Tips for Wives for Janine

When learning to fly fish, go to a women-only fishing camp or classes. There you can learn from other women skills that will be slightly different from your husband's advice. They will work better for you, and let's face it, this will be better for your relationship with your husband in the long run. There will be plenty of time on the water together, so do not rely on your husband to teach you or put him in the role of helping you with all of it.

We cross a river differently than men. Because we hold our weight in different places on our body, women tend to float, so we can lose our footing easily if the current is swift. Cross the river together; holding onto wading belts is the best.

Women will cast differently from men. Yes, women typically pick up casting more quickly. However, the rod that will fit you in weight and action may be different from your husband's. Choosing a fly rod is a lot like picking golf clubs to fit your swing. Try out fly rods at your local fly shop and see which one fits. In other words, pick your rod; don't just use an extra rod that you or your husband may have. You will also have more confidence with your rod on the water.

Fly fishing Packs – vests, sling or lumbar? Take your time to pick what you would like to have holding all your gear while you are on the stream. Honestly, it's a lot like picking a purse. There are pros and cons to each choice but find one that fits you in size and with weight from the gear. You will be glad you took the time to choose.

Do not put too much pressure on yourself. Try things out and enjoy the experience. Our lives are about accomplishing things and accomplishing them now. Don't worry about catching or not catching fish. Learn from each time on the water and look around. Trout live in beautiful places.

Author's Note from Rick – The greatest tip Jim and Janine ever taught me was how to cross a rushing river. They took me out in Colorado, and I immediately waded to the center of the river. When it came time to move to a new location, I could not. The water was rushing so hard around me, I could not move. Terrified about falling (something I am prone to do), Jim instructed me how to shuffle my feet very slowly, making sure each step had a solid footing before taking the next – all the while keeping my profile sideways to the current in order to cut down on resistance.

Follow Jim and Janine on the Facebook Page dedicated to couples flyfishing: https://www.facebook.com/groups/900021240595894

AND NOW … A LITTLE HELP FROM OUR FRIENDS

Authors' note – Couples fishing can be more than husband and wife.

Mother-Son Bonding over Walleye

My son, Logan, called me on Mother's Day and said, "The fish are biting. Come meet me at the boat landing." Of course, I heeded the call. As any serious fisher knows, you do not usually like to share a hot spot, but Logan was willing to be my guide and even let his father come along. We were barely off the landing and the smallmouth bass were biting like mad. We always catch and release those. Then suddenly, the tasty walleye started biting on the bass bait. I caught so many fish that day; I was tired! That is one good day of fishing. The best part of all is that, of course, because of Mother's Day luck, mine were the biggest. Guess whose fish is on the wall at our cabin? Yep, mine. Happy Mother's Day!

—Amy Bliss
Madison, Wisconsin

Wade a Minute!
By Ricky Wilson © 2021

The summer of 2007, from my steep fishing experience, will always be a concrete benchmark in my life as maybe the pinnacle of all the right things coming together in one place. That year will forever be the most fun I have ever had in any sport in which I have participated. That one place was Lake Cumberland, Kentucky, and its tailwaters. In early March at age 53, I was coming off my second lower-back surgery in six months, during which time I feared that my fishing days were probably well behind me. The fact that I was a predominantly stream-wade fisher added to those fears. The agility that it takes to keep one's balance in rock-strewn creeks with varying degrees of current and depth, of course, made me skeptical of ever returning to the style of fishing I loved most.

I cut my teeth fishing for smallmouth bass back in the early '70s wading the South Fork of the Licking River near towns along her banks with names like Berry, Boyd, and Morgan. It's funny when you think about and strategize on which lures will work best the day before I'd make that forty-mile journey (usually with a fellow wader) but was soon to realize that the decision about where to fish specifically was always about which part of that river was luring me to her rocky shoals and swirling semi-clear currents. In those days and especially those nights before, the anxiety would sometimes make it hard to get a full night's sleep.

A decade earlier as a kid I was heavily influenced by a TV fishing show known as "The Flying Fisher," starring an old feller named Gadabout Gaddis. He made me fall in love with fly fishing in particular, and with flying a plane, it seemed to any location he desired. As a family, my parents took us camping to some of those same places Gadabout fished. At 12 and 13 years old, yes, I had the luxury of fly fishing with my own fly rod and reel on the Yellowstone River and others like it while out west. It turns out, though, fly fishing was a bit too cumbersome in Northern

Kentucky on the streams and ponds of my youth because of weeds and cattails and overhanging tree branches. Eventually, I fell in love with Spincast equipment, specifically the Mitchel 308 reel and an Eagle Claw five-and-a-half-foot-long yellow rod. This more compact casting method seemed to be more like the right tool for the job.

Thru the '80s and '90s, The Licking River and her tributaries and creeks became my playground with varying partners in those early years and then by myself in the latter. I'd start every year on the main Licking in early February fishing for crappie at the mouth of Cruises Creek and below her riffle on the Licking. In the spring, I'd fish for smallies and white bass down on Willow Creek att DeMossville. Then I hooked smallies all summer into the fall on the South Fork.

During that period of my career, the '80s, the best fishing partner I ever had was a guy named Joe Lorenzen. Sound familiar? He was the father of the famous record-breaking quarterback for University of Kentucky football team and Super Bowl champion backup QB Jared Lorenzen, who tragically passed in July 2019. The reason I announce that Joe was the best is simple. 1) He always caught more fish than me and 2) not only did he catch more, he'd put his keepers on a stringer, lug them around all day, and then he'd give them to me after we'd walk back to the car… what a guy!

In the late '90s, I took on a project that distracted me from my wade fishing regimen for a while. Tonna (my wife of 46 years) and I moved from the overcrowded confines of Northern Kentucky to a very remote, rural location. In '98, we purchased a small cabin in Robertson County, KentuckyKY. We were drawn to this location because the cabin was above the 100-year flood plain yet only 70 yards fromThe Licking River. The cool thing about it was that it offered so many options that I had been missing since I was a kid living in rural Boone County, not to mention more opportunities to fish. However, we spent the next three years, through the sweat and blood of our own brow, enlarging, plumbing, putting in a septic system, a well, and an all-glass front with a 220-degree view of a river that runs directly through our front yard!

The wear and tear that a project of that magnitude took on the both of us at that stage of our collective lives, me 44 at the time, combined with driving 75 miles each way and her 50 to get to and from work left little energy for fishing. This only increased my yearning to get back to it though. By '05, I was starting to have serious back and neck problems. Fishing was starting to fade from my rear-view mirror and was truly out of sight by late winter of '06.

I took early retirement at age 52 after receiving MRI results that indicated I needed surgery on my lumbar spine, which I had in June of '06 and again in March of '07. My recovery was both long and painful. What was almost as painful was the depression that came when I considered I was here at last, relatively young at 52, living in our dream home and location, with a very healthy clean river bordering our front yard and two of its largest tributaries only 15 minutes in both directions, all three teaming with fish – and I wasn't able to enjoy it

Luckily, I had traded with a couple of guys I worked with and ended up with ancient but sound 15-foot jon boat and an older 15 hp Johnson, pull string motor, about a year before I was forced to retire. Time moved along well, as did the results of that second back surgery.

Within weeks I was becoming surprisingly agile again. I began putting single-set limb lines up and down this river and baiting each with small bluegill I'd catch while sitting on a bucket at a small pond a couple of miles from the house. It wasn't wade fishing, but it was very productive. River catfish, my wife and I soon discovered, are a very underrated delicious delicacy!

June came along, and with it, a large lump sum check from my disability claim! This was some much-needed good news. We had friends who owned a cabin down in Tennessee near Dale Hollow Lake. They offered us the key to their quaint little cabin about 200 miles from our house, and away we went, of course towing that little boat behind. After all, what smallmouth fisher does not know that since 1955, Dale Hollow holds the record for the largest smallie ever caught? , Weighing 11 pounds, 15 ounces, the fish was caught by D. L. Hayes.

After a few days, I caught only one fish over 17 inches. We decided to "pull up stakes" and head back up to Kentucky. Thirty-five miles north on US 127 had us crossing over the top of Wolf Creek Dam, which holds back the water that forms Lake Cumberland that was on the right of us as we were crossing. On the left is where the once-massive lake comes out and the Cumberland River's tailwaters begin. Way on the other side, on the left, was a road which descended to the electric generation station controlled by The Army Corps of Engineers who regulate how much water is discharged from the lake and which turns the turbines to generate the electricity.

That same road also puts you at the entrance to The National Trout Hatchery. We pulled in and were reminded by the volumes of live and primarily huge trout that just maybe we had been wasting our time down in Tennessee. We also learned about the massive seven-year effort to repair the leaks in the dam, resulting in the extremely low water level, which made the lake seem less than half full.

Tonna and I began to form a plan, a true "change of itinerary." We checked in at Grider Hill Marina hotel and put the boat in to take a cruise back to '76 falls, just because we had heard so much about it down through the years. There was a certain amount of eeriness from how little water was in the lake, though we were assured that there was good water back to the falls themselves.

With the lake drawn down, it was shocking how rocky and bare the beaches were, piles of rock everywhere. Now Tonna was almost ready to have a panic attack; she has always been an avid rock "hound" and insisted on being dropped off on some nondescript rocky shore on the way back to those falls. In no time, she was picking up by the hands full crystals and geodes and magnificent fossil beds of all sizes and shapes. With her having a ball, and me wanting to fish from the boat, I started catching keeper catfish using nothing but a whole night crawler on a medium-size hook and half-ounce sinker. We were both so preoccupied that we almost had forgotten why we came so far up Indian Creek.

I cleaned a couple of flatheads; she loaded her cache of treasures into the boat, and we finished our commitment to go to those "damn" falls we had both heard about. We were both so stoked about the prospects of both of our respective passions that we quickly "Chevy Chased" the falls and headed back to our hotel to eat and get an early start the next day.

That night I wrestled with sleep. All I could think about were those large rainbow trout swirling the waters of those shallow pools on display at the hatchery. We also learned that fishers were having a banner season catching them in the Cumberland River because of the current drought and the reservoir's water being drawn down so low. Intermittent releases of water from the dam created lower depths in the tail waters. In fact, we also learned that trout survival was dependent on the icy-cold water from the bottom of the massive lake and was the motivation for the minimal releases.

All the necessary ingredients were there for increased chances of successful fishing. Since I was a kid vacationing in Yellowstone, going after trout had dominated my thoughts. While fishing the lake the day before, we also ascertained from fellow fishers that the river was still navigable by small boats. That next morning, I got up two hours before the sun rose. Tonna decided to sleep in that day. I stopped at a BP gas/bait store out on Highway 127 and received a tutorial by the owner on what I needed to fish that river specifically.

I put in at the boat ramp nearest the towering dam, armed with tiny hooks, several Berkley PowerBait containers of various bright colors, and a dozen red worms. The local weather forecast said partly cloudy and midday highs in the upper 90s. However, at 6:30 a.m. with a thick fog on the river's icy-cold surface, it felt more like 50, and with low visibility, it was both uncomfortably cold and scary. You couldn't see forty yards ahead. There were probably ten trucks with empty trailers in the ramp's parking lot. You didn't know if one of them was coming at you or if simply by drifting, you might crash into someone anchored off just downstream. Until the sun rose above the lush green hills to the east and began to burn off the lingering fog, I began second-guessing my motives for venturing out onto unfamiliar waters.

Rick Robinson and Wade DeHate

I was cautioned back at the bait shop that from time to time, the operators who man the gates controlling the Cumberland's flow would sound an audible alarm when discharging sometimes dangerous amounts of water and the closer to the dam that boaters and fishers are, the greater the danger. The ongoing drought, however, made it less likely that significant release of waters would be a problem.

Though I did buy bait, this experience would prove mostly to be an opportunity to scout out the river's flow. The sheer beauty all around me served as a distraction, but I did make a few drops along the way. I took the advice given at the bait shop and put a pea-sized chartreuse PowerBait on the shank of the tiny hook and a snippet of red worm dangling from the barbed tip. Then I drifted along, bouncing the baited hook off the bottom. Two minutes was all it took, then BANG, my Ugly Stick rod bent, and a series of quick jerks had my eyes wide open and attention fixated as my Quantum spinning reel loaded with Spiderwire 10-poundtest began singing a beautiful melody of potential success. In my haste to win this first battle, and due to the river's low depth, my 15-horsepower motor's skeg bottomed out on a rock bar which swung the boat around as I fought what turned out to b, a surprisingly small 12-inch rainbow.

This began one of the most satisfying fishing experiences of my life. I caught several other trout over the next hour or two. It was enough to ensure a quick return after getting back to the boat ramp, then to the campground, where Tonna and breakfast were ready to go. Tonna doesn't fish, but is always up for an adventure, especially when hearing of the resplendent beauty lining the scenic river's path. While I was gone, she had made conversation with folks camping next to us who also had a boat. She had mentioned to our neighbors her love for rock hunting, geodes in particular. They informed her of a rock bar on the river just downstream from a place called Helm's Landing, where they also said was a prime place to catch trout. This was all I needed.

We followed the detailed map they had drawn for her. This was a very remote underdeveloped location. It was a mixed demographic of run-down shacks away from the river and upscale gentrified river mansions of the obviously privileged, many with concrete boat ramps. A small lodge (by reservation only) was located just before we finally arrived at the public boat ramp.

There were 8 to 10 fishers already wading, fly fishing mostly. I hadn't seen that many fly fishing since I was a kid in Yellowstone National Park. Tonna walked the rocky shoals as I joined those fishing.

Once off the shore and back in the boat, we finally found a hole deep enough to float the boat with the motor angled high and used the trolling motor to putter out to the deeper depths of the frigid stream. We made a beeline for the foretold rock bar down at a place called "the power lines," passing the occasional waders and boats anchored off, all of whom seemed to be catching fish.

As we drew nearer the rock bar and the river narrowed, the river's surface was riddled with scattered circular ripples belying the fact that trout were rising to the surface to pick off insects doomed to the appetite of the denizens below. 'Twas a very good sign!

Upon arrival there, Tonna stepped into the shallow icy waters and expressed her disbelief at just how cold that water could be in contrast to a day that was probably in the upper 90s. I threw out an anchor right there and began casting toward the main stream where the "risers" were occurring. I was having limited success. Some teenage boys wade fishing just downstream from us seemed to be catching fish on nearly every cast. My jealousy became so overwhelming that I eventually jumped out of the boat and waded over to accept any counseling these young locals might have to offer. The eldest of the three said the only thing he would suggest is to "tip your line with 3 feet of clear monofilament and try cooked whole kernel corn, as many as you can thread onto that small hook ya got on there." And I humbly accepted a small can of corn, only because they assured me that they had plenty. I said "Thank you very much," walked back to the boat, made adjustments and within two casts, was on my way to having one of the best fishing days of my life! "From the mouths of babes, wisdom comes" was never truer!

In fact, Tonna was witnessing all that I had discovered and decided that she too might like to give-'er-a-go. I handed her my rod and reel, gave her a few suggestions on basic casting, and she was off and running, catching fish on her first few casts. While I was breaking out my backup rod and reel to join her, a couple of rainbows she caught were so hard to crank in that she practically begged me to assist her, which I lovingly refused. What a day… what a day!

This was an experience of a lifetime at this beautiful Cumberland River and Lake and all the unique dynamics at play during that hot, hot summer of 2007. All those combinations of our own personal interests collided to the point that in the following weeks we made six more trips to that enchanting dreamlike atmosphere. There were times that we'd come 200 miles to get home, cut the grass, pay the bills, invite friends or relatives to our newly discovered paradise, and turn right around and head straight back down there. Some of those invited did indeed join us and were also as successful as Tonna and I,

A pattern soon developed. We'd fish the cold water of the river in the morning into the early afternoon, keep one fish per adult in the live well and take them up to the much warmer waters of the lake. Then under the blazing sun, anyone who wanted to rock hunt would. The rest would clean fish standing in -deep, soothing cool water, then maybe sit, drink an ice-cold beverage or two or three, and go trolling and pray to God that we wouldn't catch another fish.

—Ricky L Wilson
Jan. 31, 2021

"My husband loves to fish, but I tell him if I can't catch a fish in 20 minutes, I am not going."

—Angie Diedrich
Sun Prairie, Wisconsin

I have fished with my wife several times, and each time has become more enjoyable. The best tip I can provide, when you take her the first time or two, GET A GUIDE. Let the guide work with her on techniques and tactics, and you stay out of it. WHY? Two reasons. 1) Saves your marriage. She will take instruction from a 'professional' much better than you. She sees you as just her spouse, and how much do you really know? And in all honesty, the guide does know more and has dealt with many people and taught many more than you have. 2) You will enjoy the day more, too. While the guide is working with her, you can fish, enjoy the time, and not feel the pressure to help her. Finally, if possible, don't fish from a boat, but onshore. Much easier for the guide to teach and easier for you to stay away.

After a couple of times like this, it will be easier to go on your own, and she will have more confidence in her abilities.

—John Lovelace
Nashville, Tennessee

Having only fly fished a half dozen times, my favorite spot is Roaring Fork River at a spot outside Basalt, Colorado. Guides at Roaring Fork Club are the best. Christie never fly fished before, and she was a catching machine.

—Tom Caradonio
Daytona Beach, Florida

8

This Is Great! How Do I Teach My Kids Or Grandkids?

"Many of the most highly publicized events of my presidency are not nearly as memorable or significant in my life as fishing with my daddy."

—Jimmy Carter

Introduction to Chapter by Wade DeHate
"Cane poles on the tailgate Bobbers blowin' in the wind."
—*Almost Home by* Craig Morgan and Kerry Kurt Phillips

Probably most of us can remember with whom and where they (first) got to go fishing, and it is not always the biological mother or father being the mentor for a young boy or girl. In both past and present times, the grandparents, aunts and uncles, brother and sisters, or extended family are often teaching us the most about things like fishing.

I was the youngest of four children, and my father and mother seemed to be working all the time to make ends meet. It was my brother, Roy, who took me fishing. He is ten years older than I, and it still amazes me how much patience he had to let his much younger kid brother tag along so often.

Roy made sure I helped out by handing me a shovel to carry along to dig up red wigglers for our cane pole and bobber (Read: strike indicator if you're a fly fisher) rigs. Growing up in rural central Florida, there were many ponds and lakes around us, and a couple of hops on the shovel in any of the surrounding moist soil rewarded us with a coffee can full of worms and mud.

Many times, the hard lessons for a kid are the ones most remembered. On one occasion, one of my brother's high school buddies came along with us on a hot summer day to a bluegill pond near our home. We were all baiting our hooks with the best wiggler we could find when our companion accidentally shuffled across my monofilament line as my cane pole lay on the shore. If you've ever buried a barbed hook deep in your thumb, you can imagine the terror in my six-year-old mind at the time. This was not improved by the medical options both older boys offered of a) pushing the hook through and out another hole in my skin to get past the barb, or b) yanking it back out the same way it got there. By the way, both options suck, and as a kid or adult, you just want it to be over, and needless to say, the fishing trip was also.

As a result of letting me "tag along," my brother and I fished and hunted together many times over my childhood years before and after he left home to pursue an outdoor occupation, eventually going to work for the Florida Game and Fish Commission as a game warden in central and south Florida. We still fish and hunt whenever we can get together, and I owe much of my love for the outdoors to his patience and understanding.

Tips For Instructing Kids
1. **Start by Nurturing a Love for Fishing** – In the day and age when kids are easily distracted by everything from soccer practice to game devices, building a love of nature is not necessarily a priority for them. Understanding their impatience may cut a given fishing trip short, take

your kids and grandkids with you early. Even if they do not fish on the first trip, they will soon see you catching fish and want to participate.

2. **Put a Cane Pole in their Hands First** – Both of the authors started fishing as kids with cane poles, worms, and bobbers. A kid does not care if they have the latest equipment in their hands. They just want to catch a fish. And, let's be honest here, fishing on a lake with a cane pole is not too much different from tossing a fly – except you are likely to catch more fish with the cane pole.

3. **Start with Easier Fish to Catch** – Here again, both authors remember their first catch as bluegill. While we say start small, one of our favorite pictures on social media is a very young child holding a massive rainbow caught on his Spider-Man fishing rod and reel setup.

4. **Patience Builds Confidence** – When you are out with the kids, remember the outing is about them – not you. Be patient with them. Talk to them about what they are doing and why. Say a serenity prayer rather than a cuss word. And be ready to end the outing sooner than you would do so on your own.

5. **Safety** – You are not out with your "Hold my beer" fishing partners. You are out with children. Life vests, sunscreen, sunglasses, closed-toed shoes or sandals, hats are all the same for kids as you. Keeping a constant eye on them may require you to keep the group small. Have a low kid-to-adult ratio. Be ready for tears for a lost fish, snagged line, or other fishing frustration we encounter every day. Greet every teardrop with a smile and calm instruction on how to improve.

6. **Respect** – Teach them to respect nature. From picking up garbage to conservation, their future fishing depends on respecting the environment.

7. ***Boy Scout Booklet on Fly Fishing*** – Once you believe a child is ready to move up to fly gear, we suggest getting a copy of the booklet from the Boy Scouts of America for their Fly Fishing Merit Badge. Not for them. Unless they

are in Scouts, they are not going to read it. But for you. Following the booklet will give you a guide to teaching them the sport we love.
8. ***Fly Fishing for Kids (Into the Great Outdoors)*** by Tyler Dean Omoth – This is a great book for kids between the ages of 9 and 13 to read about fly fishing.
9. **Please Make the Experience Fun** – There is nothing more frustrating to us than seeing an adult yell at and scold a child while fishing. Our memories of fishing as a child are fun (well, except for the one where Wade got hooked). Fun fishing will keep them in the game.

AND NOW … A LITTLE HELP FROM OUR FRIENDS

Authors' note – Bob Long has spent eighteen years as a Fishing Instructor for Chicago's Fish'N Kids Program. His perspective is absolutely delightful.

Give yourself permission to catch fish any way you wish, even if it isn't the "best way" to catch a fish. Go fishing when your heart says you need to.

Whenever possible, get as close to the fish as you possibly can. And you can get closer more often than you will hear or frequently be taught. You will be surprised at how much more you will learn about fish and fishing when trying to cast for distance for the sake of distance isn't a concern.

Humans learning from humans is thousands of years old. Books, videos, magazines, and seminars are cool, but are poor substitutes for one-on-one, caring, compassionate, patient instruction. We cannot stand outside of ourselves and see ourselves. We may think we've got it, we may feel we've got it, we may swear we've got it. Then we get on the river with someone who knows what they are doing, and they go, "Oh my! What the heck are you doing? Where'd you learn that?"

—Bob Long
Chicago, Illinois

Six is a good age to learn. Grandpa Elmer showed me how to flick the line out to avoid the Wyoming sage that lined the stream in Turpine Meadows, Wyoming. The whitefish sucked up your mosquito pattern as the line straightened out. The sage tangled up the line as I hauled in my prize. Grandpa beamed with pride more than I did. He showed me how to clean the fish, and Grandma would bread it and fry it for me.

I refrained another year from giving my oldest grandson his first fly rod and setup…he is only 5 yet….

—Donald Deal
Pea Ridge, Arkansas

Fishing with Youngsters
By Greg Johnson © 2021

"You're grounded for the next five minutes, Max! Sit right here in this boat seat and don't move and don't fish. Think hard about what you know about fishing and a lot about how to act in a boat safely. Then, we'll put the pole back in your hands!" Those were my instructions to my four-year-old (closing in on five) . 1 grandson Max. I did not ask him to refrain from talking while being grounded; this would have been an impossible expectation, so there was no need to go there.

But Grandpa, you ask, why are you being so harsh on this sweet little boy? He needs to have fun, catch fish, and grandpas need to be patient. Well, wait for it; here is the rest of the story. Teaching grandkids to fish. This is great?

My son Ryan, Max, and I decided to head for Cave Run Lake in east Kentucky well before daylight on a beautiful mid-summer day with predicted temperatures in the 90s. We intended to start fishing at daylight and be off the water about noon, once the temperatures started to soar. This would be Max's third trip to Cave Run Lake that summer. I quickly learned safely boat fishing with youngsters is a two-adult activity. There are too many moving parts, such as keeping a life jacket on the youngster, feeding the youngster (top priority), untangling line, teachable moments, and endlessly answering a barrage of questions. So, having Ryan along was great; Max loves his Uncle Ryan and I love fishing with my son and grandson.

Ryan is always up for a fishing trip with his nephew. Ryan is very patient, a good virtue in today's law enforcement personnel and when fishing with kids. A sergeant and canine officer with University of Kentucky (UK) police, his training and work experience put him in a variety of situations with college students, the general public, game-day shenanigans of tailgaters, keeping his canine trained, and coordination with other law enforcement agencies in Lexington, Kentucky. When attending a UK basketball or football home game you can trust Ryan, and his canine Junior, have been there early to make sure the stadium is safe. A tall, lean man, who has fished all his life, he is a great role model for Max.

Max is a typical 4-year-old. Full of energy with a lean build and brown closely cropped hair, he is intelligent, inquisitive, intensely curious, a conversationalist, and he loves to fish. His gift for gab is already working well for him, as any fisher must have the ability to tell stories about their fishing adventures and stretch the truth when explaining the day's catch. Many a bluegill caught by him has become a state record by the end of the day when telling his story. He loves to fish.

To me, the result of teaching grandkids to fish is helping them acquire a passion for fishing. It's something they can do the rest of their lives, no matter what kind of athlete they may or not be, no matter what kind of job they may have, no matter the circumstance, fishing is always there. Your first goals for them should always include making sure they are having fun, learning safety, and most importantly, showing them you care!

I like to start grandchildren out as young as possible, allowing them to observe and hold the daily catch brought home from a fishing trip. Even as early as one year old, they love to touch fish, and marvel at a fish's colors, the texture of their skin, and the beauty and grace of one of God's greatest creations. I have found setting up a simple aquarium of easy-to-care-for tropical fish engrains a wonderment of fish movement and feeding that children never forget.

At about two years old, although all children mature differently, I give them their first fishing pole. To keep things simple, I like the little kids' poles most stores carry. With a closed-face spin-cast reel with about two feet of pole, they are cheap and easily replaceable (the first one you buy will not be the last), but hold up well to a toddler's torture, and are easy to use. I replace the monofilament line that comes with the reel with braided line. As the youngsters use these reel and rod combinations, they will always be tangling the line. It will be your duty to untangle, which you may not accomplish. I have found braided fishing line tangles less and is easier to untangle. At the end of the line is attached a casting practice plug. The game for the grandchild is to learn to cast and cast accurately. I place a five-gallon bucket in the yard and have them try to land their plug into the bucket. It's surprising from two years old to four years old how good they become at this. It's a game for them, they are having fun, and they know you care. Plus, they are developing casting skills needed for those first fishing trips.

At this point, one should mention the tackle box. First and foremost, keep yours out of reach unless you supervise the child's investigation of it. Grandkids will be in that tackle box in a heartbeat, and once done, you won't recognize the insides. Consider yourself lucky if they don't get a fish hook embedded in their finger, plus your most expensive gear seems to go missing first. Get them one of their own that's small and simple that where they can keep bobbers and other non-hook tackle. I have found it often doubles as a bug box or a critter box, having dug more than one hapless toad out of them.

Now let's take them fishing. Here the fun really begins for them and you. I prefer a small lake or stream. Again, preferably, I like to make this a two-adult trip with the grandkid, but it doesn't have to be. Expecting a three- or four-year-old to just sit on the bank of a lake or stream for more than five minutes of fishing is not logical. Two adults afford attention to grandchildren for both fishing and investigating nature, such as picking dandelions, catching crickets, throwing rocks in the water, answering a barrage of questions, and the list goes on. In Kentucky, we are blessed with a program cooperatively run by the Department of Kentucky Fish and Wildlife Resources (KFW) and local units of government called Fishing In Neighborhoods (FIN).

The FIN program is designed to make readily available small lakes in and around Kentucky counties and towns accessible for fishing from the bank. Usually, the property is owned by local governments, and the fish and lake are managed by KFW. This includes regular stocking of catfish, bluegill, bass, and trout. Just search your computer or phone for FIN program KY, and it should come up with a list of counties and lakes near you, plus the stocking schedule. These are perfect settings to start the grandkids out fishing, and most even have small playgrounds and other activities for youngsters.

I like the White Hall Park just north of Richmond, Kentucky, near the White Hall State Shrine. It has a two-acre lake that KFW keeps well-stocked, a picnic pavilion, restrooms, a playground, and blacktopped walking and biking trails. Max loves it, and this is where we started him fishing.

Max's first fishing trip was at age 3 and a half for catfish at White Hall FIN. Fairly easy to catch, and large enough to get excited about, catfish are a prefect fish to start a youngster on. Other fish, such as bluegill or crappie are also everyone's favorite for starting youngsters, and the FIN lakes have those as well.

We baited up with my strawberry chicken, a recipe of chicken breast cut into 1-inch chunks and marinated in a slurry of strawberry Jell-O mix (strawberry Kool-Aid works as well), and garlic. It's easy to put on a hook, and the catfish love it. Grandkids love helping make strawberry chicken, and letting them help usually results in a mess, but they make the connection between bait and fish. In other words, it's now their bait that they made catching their fish. Remember, it has to be fun!

For Max and his storytelling and conversation attributes, White Hall FIN Lake is the perfect setting. Sitting along the bank fishing are other anglers 10 or 15 feet from us and well within talking distance. He will spend as much time visiting these other fishers as he will fishing. When he catches a catfish, he must parade it up and down the bank, sharing it with his newfound friends and telling his story. It isn't unusual to find him netting someone else's fish for them. These are attributes I contend need to be honed in any developing fisher. The gift of gab, helping others, and the ability to develop relationships are essential in fishing and in life.

Now back to Cave Run Lake. A beautiful highland reservoir built by the U.S. Army Corps of Engineers for flood control in 1973. The dam impounded the Licking River, widely known as a "muskie" river. The lake is now known as the muskie capital of the East, thanks to KFW's good management and stocking efforts. It also has good fishing for bass, catfish, sunfish, and our quarry for the day, crappie. The lake is almost surrounded by land owned by the U.S. Forest Service, a picturesque setting for fishing, boating, and camping.

Crappie are a good fish on which to let youngsters try their newly honed fishing skills. For grandkids, I rig up a slip sinker rig with a minnow for bait (crappie love minnows). This allows the minnow to slide down to the correct depth for fish, is easy to cast, and has a bobber easily seen. When the bobber disappears, the youngster reels in his prize. Using live minnows for bait is an added plus for your grandchild. They marvel at watching minnows in the minnow bucket, they may attempt to get them out and wash them, they are not as messy as a carton of worms, but in short, the minnows can serve to be a good babysitter when needed.

This was Max's third fishing trip to Cave Run Lake, so he knew the drill. He knew the safety precautions of wearing sunblock to protect his skin, keeping his life jacket on at all times, and watching where he was moving his fishing pole and, more importantly, the attached hook. He was getting good at casting his line out, reeling in his fish, and grabbing his fish in the mouth to hold it to remove the hook. Occasionally, he needed help with all of that, and as a grandparent, or uncle, or guardian, help as much as needed, still teaching, but making sure the youngster knows you care. With Max, the conversation never ends, so when he gets quiet, one needs to be on guard. I find this to be true with my youngest grandson Myles, just now turning two. When he is quiet, it is the quiet before the storm.

Max was quiet, and when he broke the silence, he had a confession. Maybe it's because I had just reeled in a crappie and needed to measure it to see if it was a keeper. Cave Run Lake does not have a length limit for crappie, but we never keep one unless it is 10 inches long or longer. It's our self-imposed rule, but crappie much shorter do not have a lot of table fare on their bones and growing to 10 inches affords them to become much bigger fish. We have a crappie easy checker that one drops the fish into headfirst to quickly check the length. They cost about eight dollars at bait shops. If the fish is too short, just drop him right back into the water. We keep two of these measurement tools, one in the back of the boat and one in the front.

For whatever reason, Max decided to wash the one I was using. In dipping it in the water, it slipped from his hands and sank to the bottom of Cave Run Lake. This was his confession, and he felt terrible. He knew I was unhappy as horsing around in the boat, no matter what the intention, can lead to unsafe results. Tolerating much of it can lead to more serious consequences. I let him know my displeasure, and that it cost eight dollars (he is beginning to understand monetary policy at an early age), and the rules about playing in the boat. Good enough for now, and he got it. We got his pole baited up, and he went back to fishing.

Within seconds he had a fish on, he was reeling it in, and we could see it would be under our ten-inch length limit. "Max, when you get that fish reeled in, just wait, and Ryan or I will help you get it off the hook; it looks like it swallowed the hook." "Ok," replied Max. I was just thinking how proud I was of him, at almost five years old, and he was starting to fish pretty much on his own. Just then, a fish hit my jig and I noticed Ryan had one on too! Looked like the bite was picking up. Max's fish was in, and as I was reeling mine in, I heard Ryan exclaim, "Max, what are you doing?" I looked over, and for whatever reason, Max was slinging his fish against the side of the boat. He said he was trying to get the hook out, not a good idea, and he knew better. By the time I got my fish in, Max had his line in a complete bird's nest of tangles. It would take a while to untangle it, or maybe just cut it and re-rig. By any means, it was a mess. Fishing with grandkids makes you laugh and smile one minute, and the next minute use great patience to work through whatever tangle they have.

My frustration got the best of me, a standard rule not to be broken when fishing with grandkids. He knew he messed up, apologized, and gave me that long sorry-looking face that's hard to resist. I told him we probably could have gotten the hook out of his fish, the fish would have lived, and we could have returned it to the water.

Once we got all the fish in and re-rigged Max's pole, we talked about what should have been done the next time that happens. These are valuable lessons for a grandchild, and they will learn from these as much as the exciting moments. But, in a boat, on the water, rules always need to be followed for safety reasons if none other.

We had a limit of nice-sized crappie. Everyone put the chaotic incident behind, and the success stories were already being told, reminiscing the day. As we boated up on the campground beach area, we noticed a tragedy that had unfolded just minutes earlier. A jet ski was totally engulfed in flames. The fiberglass hull looked like a marshmallow on fire on the end of a campfire stick. The fire department was already on the scene, and it looked like all passengers had made it to shore. It was another valuable life lesson for my grandson, and quite frankly, for Ryan and me.

Such are the lessons and adventures teaching grandkids to fish. In the process, I think grandparents learn as much as the grandkids do. Every day is a learning experience. I'd rather fish with Max as anyone on earth. His brother Myles is right behind him. Having them both in the boat will be even more challenging and rewarding. Lifelong lessons are learned while fishing and lifelong memories are forged.

We are very fortunate in Kentucky to have the opportunities the Kentucky Fish and Wildlife Department affords through various recruit, retain, and re-engage programs. Their three conservation camps in eastern, southcentral, and western Kentucky allow more than 5,000 youngsters per year to attend and learn how to fish, swim, hunt, camp, boat, outdoor safety, archery, and more. Their Hook to Cook program also affords beginning lessons in how to fish and prepare your catch into a tasty meal. If those needing resources to help get their grandkids or nieces and nephews engaged in fishing, or you need to begin yourself to teach your youngsters, look no further than the KFW programs and their camps.

—Greg "Gravity" Johnson
Kentucky

Authors' note – Greg Johnson is the former Commissioner of Kentucky's Department of Fish and Wildlife. And, although separated by distance from his native Wasco, Illinois, he is now a true Kentuckian.

I remember as a little girl of maybe 5, my dad, Bucky Robinson, took us fishing with him and my mom Imogene on a canal near the Everglades in Florida. They had made fishing poles for my sister, Ruthie, and me. They tied one end of a fishing line to a stick and tied the other end to a clothespin. In the clothespin, they clipped some bread. Needless to say, we caught nothing. What fish would be interested in bread on a clothespin? Dad had the right gear, but I don't remember if he caught anything that day. I am sure fish were shaking their heads saying, "I don't think so!" Funny, but it is one of the highlights of my childhood memories.

—Claudia Singleton
Griffin, Georgia

Authors' note – Bucky always said, "The fishing was excellent. Catching stunk."

I like to use a nymph fly or a weighted fly and some type of little bobber for the young beginner. I then take a three-quarter-inch piece of hollow fly line and have it about a foot and a half to 2 feet above my fly floating down. It is so easy to learn because when the fish takes the fly, the indicator you're watching will stop or disappear like a little bobber. And that's when you get to lift the rod tip, set the hook say, "Good Morning, Mister Trout."

I have practiced catch and release for 34 years. And "Good morning, Mr. Trout" is very important to catching trout. Saying that keeps a fly fisher from jerking the fly out of their mouth before one can set the hook. Also, please use barbless hooks. Tight lines never lost a fish on…

—Peter Vanooyen
Star City, West Virginia

9

REFERENCES

(aka Wade and Rick's favorite books, movies, Facebook pages, and tunes)

"*Fish and whistle, Whistle and fish, Eat everything, They put on your dish.*"

—John Prine

There is no lack of instructional material available to fly fishers. With the internet full of podcasts, YouTube videos, Facebook pages and the like, your options are as plentiful as the search terms you use. Of course, the number of resources can be overwhelming to the beginner. We love fishing. Thus, our daily Facebook posts are filled with pictures of fish. Our bookshelves are filled with books about fish. However, where we fish, we never expect to snag a 34-inch fish someone caught in a country you cannot seem to find on a map. Do not fret. Join the pages and simply enjoy the scene of someone landing their personal best to make your day better.

In this chapter are some ideas of books, pages, and podcasts to check out.

FISH FICTION (Aren't all fish stories fiction?)
- *A River Runs Through It and Other Stories and Other Stories* by Norman Maclean, University of Chicago Press, 0226500608 (1989)
- *Alligator Alley* by Rick Robinson, Headline Books, 0938467654 (2013)
- *The Old Man and the Sea* by Ernest Hemingway

FISHING NONFICTION
- **John Gierach, prolific fly fishing author**—http://www.simonandschusterpublishing.com/john-gierach/index.html
- *I Never Met an Idiot on the River* by Henry Winkler, Insight Editions, 1608870200 (2011)

FISHING INSTRUCTIONAL
- *101 Knots*—https://www.101knots.com/
- *Fishing the Big Three* by Ted Williams with John Underwood, Simon and Schuster, 0671244000 (1982)
- *Fly Fishing for Kids (Into the Great Outdoors)* by Tyler Dean Omoth,
- **Flyfishing for Trout on the Cumberland River from a Drift Boat** https://www.youtube.com/watch?v=2MS4MEN7VCc
- **Fly Fishers International**—https://flyfishersinternational.org/
- **Fly Fishing Merit Badge Pamphlet**, by the Boy Scouts of America (2020)
- **Global Fly Fisher**—https://globalflyfisher.com
- **How to Handle Fish the Right Way**—https://fishuntamed.com/how-to-handle-fish/
- **Ívar's Fly Workshop**—https://www.youtube.com/channel/UCtTAi5I0OBd6JsyYdVfvBbw
- **Orvis Fly Fishing Learning Center**—https://howtoflyfish.orvis.com/
- **Project Healing Waters Fly Fishing, Inc.**—https://projecthealingwaters.org/

- **Six Women Who Are Revolutionizing the World of Fly Fishing** https://www.blueridgeoutdoors.com/fly-fishing/fly-females-meet-six-women-revolutionizing-the-sport/
- **Tenkara USA**—https://www.youtube.com/channel/UCcTgnl5MvFJ9t95LD_C2H9g
- **The Little Red Book of Fly Fishing,** by Kirk Deeter and Charlie Meyers, Skyhorse, 1602399816 (2010)
- **The New Fly Fisher**—https://www.youtube.com/channel/UCk2DJldSE7hhQTU8rjNHeYw
- ***The Orvis Guide to Beginning Fly Fishing: 101 Tips for the Absolute Beginner*** by Tom Rosenbauer (Orvis Guides), Skyhorse Illustrated, 1602393230 (2009). **For that matter, once you've gotten through 101 Tips, read everything written by Tom Rosenbauer.**
- **Trout Unlimited**—https://www.tu.org/

FACEBOOK PAGES

We have discovered Facebook has a seemingly limitless number of pages and groups related to fly fishing. These are a lot more fun than the ones where people espouse political views. We love looking at the smiles of people holding a fish far more than the frowns of people riding the political divide. Here are a few fishing pages to search for the next time you are fishin' around on Facebook. Once you have searched for a few, others will start to pop up in your suggestion box. And wait until the first time someone posts a cool comment on the photo of a fish you netted or a fly you tied.

- **Fly Fishers International**
- **Fly Fishing Addicts**
- **GoneFly - VIP**
- **Northern Kentucky Fly Fishers**
- **Project Healing Waters Fly Fishing, Inc**
- **Stupid Simple Fly Tying**
- **Trout Unlimited**
- **Worldwide Trout Anglers**

TELEVISION, PODCASTS, AND MOVIES ABOUT FISHING

- **A River Runs Through It,** Columbia Pictures (1992)
- **Fishing with John** (an old TV show now found on YouTube)
- **The New Fly Fisher Channel on YouTube**
- **The Quiet Man**, Republic Pictures (1952) *Authors' note* – There is good fly fishing instruction in The Quiet Man when Maureen O'Hara tells the priest with a prize trout on his line, "Keep his head up, you fool."

RICK AND WADE'S FISHING PLAYLIST

Sometimes a drift is all about the silence. Other times we want some tunes to accompany our efforts. We never play tunes when fishing a shoreline around others. They may well have an earworm of **Baby Shark** stuck in their heads, and we do not want to break their Karma. But sometimes, a good playlist can enhance the moment. If you do not already have a playlist set up, here are a few tunes to consider.

- *A Pirate Looks at Forty,* Jimmy Buffett
- *A Country Boy Can Survive*, Hank Williams, Jr.
- *Almost Home*, Craig Morgan and Kurt Kerry Philips
- *Bad Day Fishing*, Billy Currington
- *Choctaw Bingo*, James McMurtry
- *Cletus Take the Reel*, Tim Hawkins
- *Fish and Whistle* and *Lake Marie*, John Prine (or anything by Prine, for that matter)
- *Fishin' In the Dark*, Nitty Gritty Dirt Band
- *Fishin' Song*, Stringbean
- *Five Pound Bass*, Robert Earl Keene
- *Gone Fishing*, Bing Crosby and Louis Armstrong
- *Huntin', Fishin' And Lovin' Every Day*, Luke Bryan
- *Jerry Clower From Yazoo, Mississippi*, Jerry Clower
- *Just Fishin'*, Trace Adkins
- *It's My Lazy Day,* Merle Haggard and Willie Nelson
- *Mutineer*, Warren Zevon

- ***Shiver Me Timbers***, Tom Waits
- ***Sittin' Here Wishin' That I Could Go Fishin'***, Alton Jones
- ***Sittin' On the Dock Of The Bay***, Otis Redding
- ***Talking Fishing Blues***, Woody Guthrie
- ***The Back Porch***, Ashley McBride
- ***Walleye Willie*** and ***What The Hell Ya Got Against Fish***, Pat Daily

Acknowledgments

"Fishing is a delusion entirely surrounded by liars in old clothes."

—Don Marquis

On New Year's Eve 2020, my future daughter-in-law, Alexx Rouse, somehow took an interest in a book on fishing and made the dad jokes in my introduction acceptable to a much wider audience of readers. She is one of the best writers I know and has made this a more compelling, readable, and funnier book. A red-lined version showing her changes is available on the dark web. The Great Christmas Ornament Drop is forever forgiven. Thank you, Alexx.

We readily admitted that we do not know everything there is to know about fly fishing. One of the beauties of this sport is that you never really learn everything. Learning how to defeat a small animal with a brain the size of a tenured politician is a constant quest. Thus, thanks to everyone who participated.

Thanks to those who offered tips and stories: Marvin Lewis, Rolla, Missouri; Charles Exline, Verona, Mississippi; John Munch, Owatonna, Minnesota; Leo Poggione, Reno, Nevada; Tom "Gator" Gaither, Ludlow, Kentucky; Johnny Dennison, Lumberton, Texas; Roy DeHate, Lake Wales, Florida; Dave Waite, Ludlow, Kentucky; Sam Keating, Jacksonville, North Carolina; Barry Carruth, Glasgow, Scotland; Greg Lumpkin, Chesterfield, Virginia; Amy Bliss, Cottage Grove, Wisconsin; Tim Smith, California, Kentucky; Joel Stansbury, Florence, Kentucky; Geert Sandman, Almelo, Netherlands; Pat Schleitweiler, Cincinnati,

Ohio; Ívar Örn Hauksson, Iceland; Mike Masciangelo (www.mmtroutadventures.com); Jack Keller, Bayview, Texas; Rocky Locey, Ithaca, New York; Russ Galloway, Westbrook, Maine; Colin DeHate, Wesley Chapel, Florida; Angie Diedrich, Sun Prairie, Wisconsin; John Lovelace, Nashville, Tennessee; Bob Long, Chicago, Illinois; Donald Deal, Pea Ridge, Arkansas; Claudia Singleton, Griffin, Georgia; Peter Vanooyen, Star City, West Virginia; Tom Caradonio, Daytona Beach, Florida; John Mocker, Union, Kentucky; Bob French, Chicago, Illinois; Melissa Moody, Salem, Arkansas; Marcus Carey, Owenton, Kentucky; Colin DeHate, Florida; and Brandon Wade and Michael Wlosinski of Cumberland Drifters.

Thanks to those who offered copyrighted essays: Spencer Durrant (www.spencerdurrant.com), Peter Collin, Mark Forster, Mark Neikirk, Mark T. Sorrels, Jack Kerley (www.jackkerley.com), Steve Champion, Jim and Janine Young, Ricky Wilson, and Greg "Gravity" Johnson and Tim Farmer (timfarmerscountrykitchen.com).

And thanks to generations of writers, authors, and conservationists whose quotes we gathered in these pages under the Fair Use Doctrine. Put our shameless use of the doctrine to good use — buy their books and read their articles: W.D. "Bill" Gaither (1942-2016); Moses; John Buchan (1875-1940); Jesus of Nazareth; Simon Peter; John D. Voelker (aka Robert Traver-1903-1991); Mark Twain (1835-1910}; Nick Lyons; Ernest Hemingway (1899-1961); Tom Rosenbauer, Norman Maclean (1902-1990}; G. E. M. Skues (1848-1949), Henry David Thoreau (1817-1862); Jerry Clower (1928-1986), Studs Terkel (1912-2008); Zane Grey (1872-1939); William Tapply (1940-2009); President Herbert Hoover (1874-1969); Harry Middleton (1949-1993); Washington Irving (1783-1859); David Stuver (www.riverfeetpress.com); Ted Williams (1918-2002); Bob Lawless; Steven Wright (www.stevenwright.com); John Gierach (www.simonandschusterpublishing.com/john-gierach/index.html); Brandon Wade and Michael Wlosinski of Cumberland Drifters; John Steinbeck (1902-1968); Harold F. Blaisdell; Tony Blair; Koos Brandt; Thomas McGuane; Reg Baird; George W. Snyder

(1780-1841); G.S. Marryat (1840-1896); Jimmy Moore; Orlando Aloysius Battista (1917-1995); Edward Abbey (1927-1989); Charles Orvis (1897-1915); Jeff Foxworthy (www.jefffoxworthy.com); Roderick Haig-Brown (1908-1976); Tom Brokaw; Leonardo Da Vinci (1452-1519); Izaak Walton (1593-1683); Rial Blaine; Clare Vanderpool; Henry Winkler; Julia Child (1912-2004); Leon Wulff (1905-1991); Mitch Hedberg (1968-2005); Jessica Maxwell; Don Marquis; Eric Barker; Eric Hayes; Kristopher M. Kuhn; R.A Ferguson; B.L. Tufts; James Lamansky, Jr.; Kevin Meyer; Luciano Chiaramonte; Don Whitney; Joshua McCormick; and President Jimmy Carter.

Thanks to Laura Bezold for providing the self-portrait painted by her father, WD "Bill" Gaither (aka Catfish Daddy-O).

A special shout-out to our wives, Robin DeHate and Linda Robinson, who never complain when we sneak off to the river.

Finally, thanks to Cathy Teets and everyone at Headline Books for making us write this book. We have had an absolute blast with this project.

About the Authors

Rick Robinson is an international award-winning author, having twice been named the International Independent Published Author of the Year. He has ranked number one in his category on Amazon and has often placed multiple books in the top 100. Rick has also been a regular humor columnist for local, national, and international publications. Rick and his wife/editor Linda live in Kentucky. Rick's other books are:

The Richard Thompson Series
- *The Maximum Contribution*
- *Sniper Bid*
- *Manifest Destiny*
- *Writ of Mandamus*
- *The Advance Man*
- *Opposition Research*

The Coming-of-Age Series for Folks Who Never Came of Age
- *Alligator Alley*
- *The Promise of Cedar Key*

Other Books
- *Landau Murphy, Jr. – From Washing Cars to Hollywood Stars*
- *Killing the Curse* – with Dennis Hetzel
- *Strange Bedfellow* (Kindle only)

Growing up learning to fish and hunt in the southeastern United States, **Wade DeHate** is an avid outdoorsman. During the mid-1980s, Wade discovered fly fishing for trout. Over thirty-five years later, he continues to hone his skills by taking the time to appreciate the many types of fishing methods local anglers use from the Caribbean to Alaska, even spending some time catching a few trout with some folks in Ireland and Scotland.

Wade's big brother was a game warden, and it influenced both his understanding of and commitment to the conservation of our natural resources. Like those who taught him, Wade passed on these conservation traits by mentoring young people during several years as a Scoutmaster, reinforcing the concept that if you want to understand something better, teach it.

Wade's thirty-one-year career as a firefighter fed his desire to influence the quality of life in his community. In the later years of his fire rescue career, he developed his technical writing skills by authoring multiple documents and components of local, state, and national codes and standards for the National Fire Protection Association (NFPA) as well as original emergency response research at the FEMA National Emergency Training Center at the U.S Fire Administration.

After retiring from the fire service in 2006, Wade and his wife, Robin, ran their own consulting company, writing emergency response plans and teaching chemical spill response around the southeastern United States. Wade and Robin have three wonderful children and now spend many of their days on the Cumberland River in Kentucky.